Hardware Int

with

RobotBASIC

- The Fundamentals -

by

John Blankenship and Samuel Mishal

Contents at a Glance

Preface_____vii
1- Introduction to I/O Ports_____1
2- Parallel Port Examples_____7
3- Serial Port Examples_____19
4- Motor Control_____31
5- Sensors_____43
6- Speech, the Human Interface_____55
7- Vision_____65
8- Communications and Control Over the Internet_81
9- What's Next_____87
A- Bit-wise Operations_____97
B- Finding Serial Port Numbers_____103
C- Utilizing the TCP and UDP Protocols_____105
D- Byte Buffers in RobotBASIC_____153
Index_____165

Table Of Contents

Preface_____vii
1- Introduction to I/O Ports_____1
    Parallel Ports_____1
    Serial Ports_____2
    Comparing Serial and Parallel_____2
    Synchronous vs. Asynchronous_____2
    Modern Ports are Serial_____3
    Converting Serial to Parallel_____4
    Virtual Serial Ports_____4
    I2C Communication_____4
    Virtual Parallel Ports_____5
    Utilizing Microcontrollers for I/O_____5
    Summary_____5
2- Parallel Port Examples_____7
    OutPort_____7
    InPort_____11
    USBmicro I/O Boards_____13
    USB Alternatives_____17
3- Serial Port Examples_____19
    Serial Communication with a Microcontroller___19
    The BS2 Controller from Parallax_____20
    Synchronizing Communication_____20
    Analyzing the Program_____21
    The BS2 Program_____22
    Connecting the Hardware_____24
    Adding an Input Port_____25
    Creating New I/O Functions_____26
    Creating Custom I/O Functions_____28
    Conclusion_____29
4- Motor Control_____31
    DC Motors_____31

Buffering the Port_____32
Controlling Speed_____32
Controlling the Direction_____33
Servomotors_____37
Microcontrollers Can Help_____38
Micro Maestro 6-Channel USB Servo Controller_38
Putting it all Together_____40
Summary_____41
5- Sensors_____43
Modern Sensor Technology_____43
Types of Sensors_____43
Bumper Sensors_____44
IR Perimeter Sensing_____44
Ranging Sensors_____46
Building a System_____49
Line Sensors_____51
An Electronic Compass_____52
Other Sensors_____53
6- Speech, the Human Interface_____55
Synthesizing Speech_____55
Voice Recognition_____56
Eliminating the Need for ENTER_____58
An Example Voice Recognition Application____59
7- Vision_____65
Inexpensive Web Cams_____65
Analyzing the Image_____65
RobotBASIC's Vision Capabilities_____66
RoboRealm_____72
Alternative Communication_____79
Summary_____80
8- Communications and Control Over the Internet_81
Internet Communication_____81
Controlling a Simulated Robot Over the Internet_82
LAN or Internet_____86
Summary_____86

9- What's Next_____**87**
   Why RobotBASIC?_____87
   A New Paradigm in Robotics_____88
   RobotBASIC's Internal Protocol_____91
   Advantages of the New Paradigm_____94
   Summary_____96
A- Bit-wise Operations_____**97**
   Logical Operations_____97
   Bit-wise Operations (2 Inputs)_____97
   Clearing Bits_____101
   Setting Bits_____101
   Toggling Bits_____101
   Unary Operations_____102
B- Finding Serial Port Numbers_____**103**
C- Utilizing the TCP and UDP Protocols_____**105**
   Scope_____106
   Some Terminology_____106
      A Local Area Network (LAN)_____107
      An Internet Service Provider (ISP)_____108
      The Internet_____108
      A Router_____108
      WiFi_____108
      Fire Wall_____109
      Dynamic Host Control Protocol (DHCP)____109
      Network Address Translator (NAT)_____109
      Socket_____109
      Transmission Control Protocol (TCP)_____110
      User Datagram Protocol (UDP)_____110
      Simple Mail Transfer Protocol (SMTP)_____111
      Remote IP Address_____111
      Local IP Address_____111
      Packet_____111
   Utilizing the UDP System:_____111
      Sending Data Using the UDP_____113
      Receiving Data Using the UDP_____114
      Checking the Socket's Status_____114

Receiver-Side Automatic Header Appending_116
Sender-Side Manual Header Appending_____116
Developing a UDP Program_____117
A Simple UDP Program_____117
Another UDP Program_____118
A More Practical UDP Example_____119
The UDP_UserIO.Bas Program_____120
The UDP_Calculator.Bas Program_____123
Suggested Improvements_____124
Utilizing the TCP System_____127
Sending Data Using the TCP_____129
Receiving Data Using the TCP_____130
Checking the Socket's Status_____131
Server-Side Automatic Header Appending__132
Sender-Side Manual Header Appending____133
Developing a TCP Program_____133
A Simple TCP Program_____133
An Improved TCP Program_____134
A More Practical TCP Program_____136
The TCPC_UserIO.Bas Program:_____137
The TCPS_Calculator.Bas Program:_____139
Suggested Improvements_____141
Selecting a Port Number_____143
Allowing Internet Throughput_____144
Configuring the Router for Port Forwarding   148
D- Byte Buffers in RobotBASIC_____153
Manipulating A Byte Buffer_____153
Putting Text and Numbers in a String Buffer____154
Putting Text and Binary Numbers in a Buffer___157
Bytes_____157
Integers and Floats_____159
Extracting Text and Numbers from a Buffer____161
Another Design_____162
An Efficient and Flexible Design_____163
Index_____**165**

Preface

Although RobotBASIC is best known for its integrated robot simulator, it also has extensive commands for interfacing with external hardware, making it very useful for many control applications in addition to robotics.

Based on the emails we receive, there are many people who want to experiment with controlling a robot and electronic hardware through a personal computer but don't have the necessary skills in electronics or programming. In this book we will show how RobotBASIC and some readily available hardware, can make the above goal easily attainable.

Modern technology has greatly reduced the complexities of communicating with, and controlling external devices. Companies like Parallax, Lynxmotion, and Pololu now provide motor controllers and sensors that make it easy to accomplish projects that would have required an engineering degree only a short time ago. Even with all these advancements though, a proper understanding of the fundamental principles can help ensure success.

This book concentrates on how RobotBASIC programs can interface with a wide variety of off-the-shelf hardware. Many options will be explored including both serial and parallel ports and how a separate microcontroller can be used to handle the low-level I/O operations. We will also show how vision, speech synthesis, voice recognition, and Internet communications can be used in your RobotBASIC programs.

The main portion of the book uses straightforward, easier examples to introduce the concepts. A series of appendices provide additional details for more intricate information.

Visit our web page to download your free copy of RobotBASIC.

www.RobotBASIC.com

Chapter 1

Introduction to I/O Ports

This chapter will introduce the various types of ports and some of the terminology associated with them. Later chapters will discuss the details of using these ports.

When you write PC programs that need to obtain sensory data from external devices and control electro-mechanical actuators such as motors, you need to understand what ports are and how to use them.

Input/Output ports can be subdivided into two general categories, serial and parallel.

Parallel Ports

Parallel ports generally make available 8 pins for either input or output or both. That means they can transfer 8 bits of I/O. Other sizes are also possible. The term *bit* originated from the phrase, *b*inary dig*it*, which simply means that each piece of information can only be one of two values. These values are typically referred to as *zero* and *one*, *on* and *off*, or *low* and *high* where low and high generally correlate to voltage levels in an electronic circuit.

When standard TTL (Transistor-Transistor Logic) circuitry is used, a low usually means zero volts and a high is typically associated with five volts. Since a bit can only take on one of two possible values, there is a threshold voltage (typically around 3 volts, but that varies by the types of electronics being used) that marks the boundary between a logical high and low.

A group of 8 bits is often referred to as a *byte*, and two associated bytes (16 bits) are often called a *word*. There is even a term for a 4-bit grouping. It is called a *nibble* (half a byte).

When a computer program needs to transfer a group of bits to or from some external hardware, a parallel port is required. This term simply means that a group of bits are transferred simultaneously - that is, in parallel.

Serial Ports

Serial ports get their name because the bits of information are transferred between devices in a serial manner - that is one bit at a time. Serial ports come in many different types, as we shall see later.

Comparing Serial and Parallel

Imagine that we have eight ping-pong balls, each labeled with a 1 or 0. Assume that together they make up a byte of information that needs to be transferred from one device to another.

A parallel port is like gathering all eight balls together and throwing them to someone. If you throw the balls one at a time, it is similar to a serial port.

One advantage of a parallel port transfer is speed. It is obviously faster to transfer all the bits at once rather than sending them sequentially. Older serial ports were asynchronous and very slow compared to today's synchronous serial ports.

Synchronous vs. Asynchronous

The most common *asynchronous* serial transfer protocol is called RS-232. The actual data being transferred is sandwiched between a *start* and *stop* bit. The start bit simply gives the receiving hardware a wake-up call, indicating that the real data is coming. The start bit is necessary in asynchronous transfers because the data can be sent without warning (thus the term asynchronous). It is the responsibility of the receiving hardware to watch for the arrival of data. Timing is of paramount importance and it is measured in relation to the time of arrival of the start bit. The stop bit does not actually indicate the end of the data, but only serves as a small delay between the transfer of each group of 8 bits to ensure that the next start bit can be distinguished from the last bit of the previous 8 bits.

A *synchronous* transfer requires two wires instead of just one used for asynchronous transfers.

> ⓘOf course, both methods also require a ground wire to complete the electrical circuit.

On synchronous transfers, the second wire carries a *clock* signal whose pulses indicate exactly when data is available on the first wire. This can make synchronous transfers much faster than asynchronous ones because this additional clocking line helps make the timing much more precise which helps in maintaining the transfer integrity and synchronization even at high clocking rates.

Modern Ports are Serial

Not long ago, all printers came with a parallel port interface operating through a 25-pin connector. Actually, the interface was composed of three separate parallel ports. One port carried the data (the codes for the letters to be printed) and a second port was used to send error information (paper out, printer off line, etc.) back to the computer. The third port was used to control the transfer, telling the printer when another piece of data was ready to be taken. RobotBASIC has commands that allow reading and writing to parallel ports, even though most computers today do not have them anymore.

Nearly all printers today communicate using a USB port, which is a synchronous serial port with well-defined standards. Because of its synchronous nature, USB ports are nearly as fast as parallel ports, but can be implemented less expensively.

Converting Serial to Parallel

When the 8 serial data bits arrive at their destination they must be reconstituted back into a byte so that it can be used in the computer (or microcontroller) as a single byte representing a number or character. Fortunately, most computers and microcontrollers provide either hardware or software modules that make it easy to send and receive serial data relieving you from having to understand all the low-level details.

Virtual Serial Ports

On the PC, the Windows OS allows a device plugged into a USB port to establish itself as a virtual serial port. Since many languages, including RobotBASIC, have commands to read and write to a serial port, this provides an easy way to communicate with many modern external devices.

Most modern PCs no longer have the original RS-232 serial ports just as they no longer have parallel ports. However, many microcontrollers and even some sensors communicate using the RS-232 standard, so robotic projects will often need a standard serial port. In order to provide compatibility, many companies provide a USB-based serial port.

I²C Communication

USB is not the only synchronous interface. Many serial memory modules, electronic compasses, and other devices useful to robotic hobbyists use the I^2C synchronous interface which has the advantage of being able to communicate bi-directionally with only two wires (in addition to the ground wire).

As with the USB interface, it is often easy to find hardware or software modules that make it easy to utilize

I²C communication without having to understand all the details of its internal low-level operation.

Virtual Parallel Ports

Just as many companies provide the necessary circuitry to create a virtual RS-232 serial port over a USB port, so there are also companies that provide circuitry to create a virtual parallel port using a USB port. One such company is USBmicro (www.USBmicro.com). Their U4x1 devices are not just parallel port replacements. They also provide many powerful features (see next chapter). RB has a suite of functions to control the U4x1 and utilize all its facilities.

Utilizing Microcontrollers for I/O

Since microcontrollers are designed for low-level control applications, they provide an easy way to add I/O operations to nearly any PC. The microcontroller can be programmed to communicate with RobotBASIC using an RS-232 interface. This allows the controller to perform the actual Input/Output operations on its I/O pins as requested through serial commands sent by RB.

You might dislike the idea of having to program a micro controller to act as your I/O interface, especially since one of the major advantages of using RobotBASIC to control a robot is that you get to program in an easy-to-use high-level language. There is an important distinction though. The program in the controller only has to handle very basic operations, so it is easy to write, and, as you will see in later chapters, the low-level program generally only has to be written once. All of the major programming, that is all of the decision-making and intelligence, can be fully implemented in RobotBASIC.

Summary

This chapter provided an overview of various types of ports and some terminology associated with them. The next chapter will provide some practical examples of how to utilize parallel ports in your RobotBASIC programs.

Chapter 2

Parallel Port Examples

In this chapter, we will look at some practical examples of how a PC's parallel I/O ports can be used to send and receive data to and from external circuitry. Let's start with the most generic I/O commands in RobotBASIC, `InPort` and `OutPort`. These commands allow you to read and write to a valid I/O port on your PC. Because of that, these commands should be considered *dangerous*. Many PC devices (such as disk drives, video cards, sound cards, and interrupt controllers, just to name a few) use I/O ports to control their operation. Inadvertently writing improper data to the wrong ports can alter your PC's operation in strange ways and potentially damage it.

OutPort

The `OutPort` command sends data to some external circuitry and requires two parameters as shown below.

```
OutPort PortNumber, Data
```

The **PortNumber** is an address that specifies which of 65,536 ports to use (addresses 0-65,535). Most port addresses are not used, and of those that are, many should only be used by the operating system or other programs that know how to use them properly. Let's see how to use `OutPort` by looking at a specific example.

Older PC's had a special 25-pin connector for interfacing with printers. Nowadays, most computers use a USB port for connecting to a printer, so spending a lot of time on the older interface would not be fruitful. It does provide a simple introductory example though, and you may have an older computer that has a parallel printer interface. This example is also valid, because the techniques shown here can be used to exchange data with *any* parallel port, not just the printer port.

> ⓘRobotBASIC can run on any version of the Windows OS from 95 to Vista and 7. If you have an older computer you can use it to do your experimentation. RB needs no installation and is easily usable from a thumb-drive or CD or even through the Internet. Using this possibility means you can make use of old PCs and you do not have to worry about damaging it.

As mentioned earlier, the old parallel port printer interface is actually composed of three different I/O ports; a data port, a control port, and a status port. The data port was an 8-bit output port used to send data to the printer. The control port was a 4-bit output port used to initialize the printer and to tell it when new data was available on the data port. The status port was a 5-bit input port that allowed the PC to receive information from the printer (indicating, for example, that the printer is busy or out of paper).

> ⓘYou can specify a hexadecimal number in RobotBASIC by using 0x. For example 0xA4B3 would mean the hexadecimal number A4B3, which is 42163 in decimal. To specify a binary number use 0%. So 0%1101 is the binary number 1101, which is 13 in decimal.

On most desktop PC's the location for these ports started at a base address of 0x378 (888 in decimal), although some PC's and laptops use a different address. The data port was located at the base address followed by the status port and then the control port. For this example we will only use the data port. Let's assume we have three LED's connected to the three least significant bits (LSB) of the port, as shown in Figure 2.1. The figure also shows the pin assignments for the 25-pin connectors used for parallel printer interfaces.

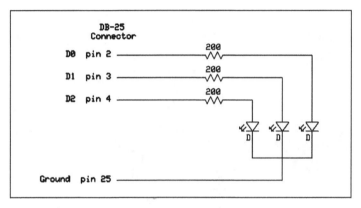

Figure 2.1: Three LEDs connected as shown allow us to see the data written to the port.

When a 1 is written to one of the port pins, that LED will light because it will have a voltage applied to its anode terminal (typically 3-4 volts) and ground to its cathode terminal. We can light all three LEDs by writing a 7 (binary 111) to the port. Let's see how to light the LED's from RobotBASIC. Look at the one-line program below.

```
OutPort 0x378, 1
```

If you run this program with the LEDs connected, the D0 LED will light. If you send a 2 (0%010) to the port, the D1 LED will light, and a 4 (0%100) will light D2. Remember, this program assumes the parallel port is at address 0x378.

If your printer port is at some other address, or if you are using a different type of parallel port entirely, you must use the appropriate address.

The short program in Figure 2.2 will cause the D0 LED to blink at a one-hertz rate.

```
while true          // loop forever
   OutPort 0x378,1  // turn the LSB on
   delay 500        // wait ½ second
   OutPort 0x378,0  // turn all bits off
   delay 500        // wait ½ second
wend
```

Figure 2.2: This program blinks the LED on the LSBit.

We can make the light appear to move back and forth from one side to the other by sending the proper data to the port. In particular, the data we need to send to the ports is shown below. If these numbers are sent over and over, the lit LED will appear to move back and forth.

Decimal	Binary
1	001
2	010
4	100
2	010

The program below shows how to accomplish this in RB. You can adjust the delays to make the light move faster or slower.

```
while true
   OutPort 0x378, 0%001
   delay 200
   OutPort 0x378, 0%010
   delay 200
   OutPort 0x378, 0%100
   delay 200
   OutPort 0x378, 2
   delay 200
wend
```

Figure 2.3: This program causes a light to move back and forth on three LEDs.

InPort

Let's see now how to input data from any PC port. The
InPort function is similar to OutPort except that it
returns the data from the designated port address. For our
next example, we will assume the circuitry in Figure 2.4 is
connected to the status port for the parallel printer interface.

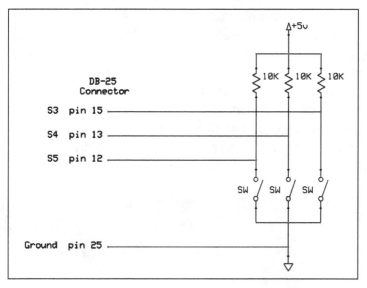

Figure 2.4: The status port can be used for input.

Each input pin is connected to a switch and a resistor.
When the switch is closed, the input line is pulled to ground
causing the input to read a zero. When the switch is open,
the pull-up resistor will pull the line towards five volts,
causing the input to read a logical one. This means that a
normally-open switch will input a zero when pressed. If
you prefer the opposite, you can use normally-closed
switches or invert the data either with hardware or in
software. It is worth mentioning, that since the printer port
connector does not have the 5-volt supply available, that a
separate 5-volt supply must be connected. Note also that
the input data lines do not enter on the least significant bits

of the port (S0 would be the LSBit). This means that the data must be shifted to the correct position after it has been obtained.

The program in Figure 2.5 obtains the data from the switches and displays it in the raw form as well as inverted and shifted. The program prevents flicker by updating the display only when a new number is obtained from the port by comparing the number obtained to the old number. Also notice how the bit-wise AND (& or bAND) is used to *clear* (or *mask out*) all the unwanted bits. This action only keeps bits marked by 1's in the number &ed with the data. Individual bits in data can be *inverted* by bit-wise exclusive-ORing them with a one. In this case, we want to invert the three least-significant bits, so we bXor (or you can use the @ operator) them with 7. If you are not familiar with bit-wise operations, refer to Appendix A.

```
d=0 \ oldData=d \Gosub DisplayData
while true
   InPort 0x379, d     // get data from port 0x379
                       // and place in d
   d = d & 0%00111000 // zero all unwanted bits
   if d<>oldData
      Gosub DisplayData
      oldData=d
   endif
wend
end
DisplayData:
   clearScr
   print "        Raw Data = ",d
   print "           binary ",bin(d,3)
   print
   print "        Shifted Data = ",d>>3
   print "           binary ",bin(d>>3,3)
   print
   print " Shifted & Inverted = ",(d>>3)bXor(7)
   print "           binary ",bin((d>>3)@7,3)
return
```

Figure 2.5: This program displays data from an input port in several forms.

Assuming you use normally open switches, then if all the switches are off (open) then each input pin will read a high or 1. Since these bits are coming in three bit positions to the left, if all bits are high, the number displayed in the raw mode would be 56 (0%000111000) and the output from the program would look like Figure 2.6.

Once the raw data is obtained from the port, it can be converted to a more desirable form by shifting it right three positions using the shift operator (>>) and inverting the desired bits by binary exclusive ORing them (@) with 1's.

```
        Raw Data = 56
          binary 111000

     Shifted Data = 7
          binary 111

Shifted and Inverted = 0
          binary 0
```

Figure 2.6: This is the display from the program in Figure 2.5 when all the switches are open.

The programs examined so far demonstrate the basic principles for dealing with I/O ports utilizing the legacy parallel printer port. If you are using an older PC, or if you have added a third party parallel port board to your PC, then these principles apply. Most modern personal computers though, communicate with external hardware with USB ports. RobotBASIC supports USB ports in multiple ways, one of which is supplying special commands for controlling products from USBmicro.

USBmicro I/O Boards
USBmicro (www.USBmicro.com) produces parallel port I/O boards that can be connected to a PC through the USB ports. The U401, shown in Figure 2.7, is fully supported by RobotBASIC with commands that communicate with a special DLL file supplied by USBmicro. We will explore some of the basic commands here, but refer to

RobotBASIC's HELP file and to USBmicro's web page documentation for complete information.

The U401 has 16 I/O lines that can be individually configured as inputs or outputs. The 16 lines are subdivided as two 8-bit ports, A and B. The required USBmicro DLL (USBm.DLL) is provided when you download the RobotBASIC zip file for a full install.

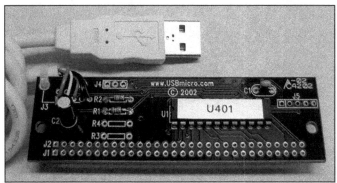

Figure 2.7: This U401 I/O board from USBmicro
is fully supported by RobotBASIC.

The output pins on a U401 board have VERY limited current capabilities, so they should be buffered with some form of buffer/driver as shown in Figure 2.8. When a logical one is applied to the inverting buffer, it provides a low output that turns on the associated LED. The rest of the circuitry is similar to what we have seen before.

There are many special RobotBASIC commands for controlling the U401. The program in Figure 2.9 demonstrates a few commands that read the switch data from Port A and send it to the LEDs on Port B. When this program is running, anytime you change the switches, you will see the results on the LEDs. Notice how commands are used to verify that both the USBmicro DLL and I/O board have been found.

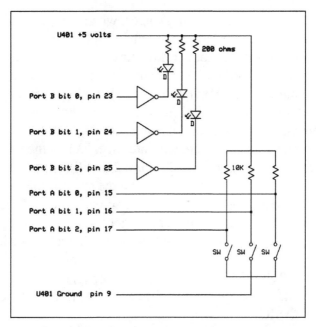

Figure 2.8: External devices are easily connected
to the U401 from USBmicro.

```
if usbm_DllSpecs() != ""
  if usbm_FindDevices()
    //---there is a device and it will be device 0
    usbm_DirectionA(0,0,0) //set port A0 to A7
                           // as inputs
    usbm_DirectionB(0,0xFF,0xFF) //set port B0 to B7
                                 //as outputs
    while true
      n = usbm_ReadA(0) // assume device 0 for
                        //both read and write
        usbm_WriteB(0,n)
      wend
    else
      print "There are no Devices"
    endif
  else
  print "The USBmicro DLL is not installed"
endif
```

Figure 2.9: This program reads switches and
mirrors their status on LEDs.

RobotBASIC has many advantages, when compared to a microcontroller, for handling control applications. Microcontrollers, no matter how powerful they are, usually have limitations when it comes to memory size, variable types, multi-dimensional arrays, and high-level language support. The program in Figure 2.10 demonstrates how the simple task of reading and writing port data can be enhanced by using a graphical interface. When the program is run, the LEDs of Figure 2.8 will reflect the state of the checkboxes on the screen and the state of the switches will show in binary on the screen. Some of the lines in our program listing are longer than the book's page width. When this happens, we use the \ character to allow the code to continue on the next line. You may enter the line into RobotBASIC using the \ character, or you may enter it all on the same line. Either way is acceptable to RobotBASIC.

These examples show only a small portion of the power of the USBmicro I/O boards. Refer to the RobotBASIC HELP file and www.USBmicro.com for more information and additional examples.

```
if usbm_DllSpecs() != ""
  if usbm_FindDevices()
    //---there is a device and we will use device 0
    usbm_DirectionA(0,0,0)   // set port A0 to A7
                             //as inputs
    usbm_DirectionB(0,0xFF,0xFF) //set port B0 to B7
                             //as outputs
    xyText 10,10,"Set Output Port Status:" \
                  ,,20,fs_Bold|fs_Underlined
    for i=0 to 2
      addcheckbox ""+i,430+20*(2-i),20," "
    next
    xyText 90,70,"Input Port Status:" \
                ,,20,fs_Bold|fs_Underlined
    rectangle 425,70,493,108
    while true
      for i=0 to 2
        if getcheckbox(""+i) // see if each box is
                            // checked
          usbm_SetBit(0,8+i) //add 8 for
                             //Port B bits
        else
          usbm_ResetBit(0,8+i)
        endif
      next
      // the following is done individually to
      // clarify each task
      n= usbm_ReadA(0)    // read the port
      n= n bXor 7         // invert lower three bits
      n= bin(n,3)         // convert to a binary
                          //     string
      // note: all the above can be done in one line
      //n = bin(usbm_ReadA(0)@ 7,3)
      xyText 437,75,n,,20,fs_Bold
    wend
  else
    print "There are no Devices"
  endif
else
  print "The USBmicro DLL is not installed"
endif
```

Figure 2.10: A Graphical Interface can enhance interaction with ports.

USB Alternatives

The USBmicro boards are one of the easiest ways of connecting to the real world though a USB interface.

Moreover, these boards provide much more than just normal on/off I/O. The boards allow you to control stepper motors with ease and to carry out I^2C, SPI and 1-Wire synchronous serial communications and to control an LCD. The U451 also allows you to switch high-voltage-high-current relays (240V 10A). These properties can be quite useful in many projects.

In the next chapter we will see another method for interfacing with a USB port.

Chapter 3

Serial Port Examples

Based on comments in Chapter 1 and 2, you might suspect that microcontrollers should not be used for control application. To the contrary, there are many applications, such as microwave ovens and programmable thermostats, for example, that are ideal candidates for the low-cost computing power of a micro controller. Furthermore, hardware applications that need the features of a high-level language can often be implemented more easily by utilizing a microcontroller for handling the low-level I/O tasks.

Serial Communication with a Microcontroller
Utilizing microcontrollers in conjunction with RobotBASIC provides many advantages. To demonstrate this point, we will use serial commands to interface RB with the Basic Stamp 2 microcontroller (BS2) from Parallax so that it acts as parallel input and output ports for us. This example provides a practical application for introducing serial I/O operations.

It is important to realize that nearly any micro controller can be used in this way as long as it has the features needed (such as number of I/O lines). For this text we will concentrate on the Parallax BS2 controller using the Board Of Education (BOE) carrier board shown in Figure 3.1.

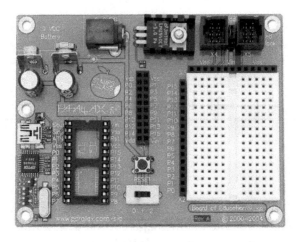

Figure 3.1: This Board Of Education carrier
board supports the Basic Stamp 2 (BS2).

The BS2 Controller from Parallax

The BS2 has 16 I/O lines and we will use two of them for
serial communication with RobotBASIC. That leaves 14
lines free, but we will only use 8 lines and implement them
as one 4-bit input and one 4-bit output port.

The BS2 has commands for sending and receiving serial
data, but there are some limitations that have to be
considered. For example, if the BS2 is expecting to
receive serial data, it must abandon all other tasks and wait
patiently for the data to arrive. While this may seem like a
huge problem, it can be easily addressed by creating a
protocol between RobotBASIC and the BS2 that
synchronizes the data transfer.

This synchronization simply means that the two systems
take turns sending and receiving data. An example will
clarify this idea.

Synchronizing Communication

Assume for a moment, that we will program the BS2 to
receive 4 bits of output-data from RobotBASIC. When the
data is received, the micro controller's job will be to send

that data to four of the controller's I/O pins, thus allowing the controller to serve as a 4-bit output port for RB. As long as RB does not send data too quickly, this should work fine. If RB sends new output data before the controller is ready to receive it though, the new data could be lost.

We can easily solve this problem by having the controller let RB know when it is again ready for data. The controller does this by sending some sort of reply back to RB. For this example, we will assume the controller sends back the same data it received.

Since both RobotBASIC and the controller are always waiting for communication from each other, it becomes impossible for either to be busy when the other is trying to correspond. Figure 3.2 shows a simple program that flashes four LEDs connected to the output port on the BS2. It also provides a simple example for us to learn about the RB serial I/O commands. Obviously, the BS2 controller will have to be connected to a serial port on the PC, and programmed to respond appropriately. We will do this shortly.

Analyzing the Program

Let's examine the program in Figure 3.2 to see how it works. The first line in the program tells RobotBASIC which serial port to use. See Appendix B to find out how to determine what the port number should be. For now, just assume that the BS2 is connected to port 33. Once the SetCommPort executes, all subsequent serial commands will utilize the specified port.

```
SetCommPort 33
while true
  SerialOut 15 // all 4 bits high
  gosub WaitForReply
  delay 500    // on for ½ second
  SerialOut 0  // all 4 bits low
  gosub WaitForReply
  delay 500 // off for ½ second
wend
end

WaitForReply:
  repeat
    CheckSerBuffer numBytes
  until numBytes=1
  SerIn d
return
```

Figure 3.2: This program sends data serially to a BS2, and
waits for a response before continuing.

Next, an endless `while`-loop causes the main portion of
the program to continually repeat itself. Inside the loop, a
`SerialOut` command sends out the number 15 (binary
1111) to turn on all four LEDs. Later in the loop the
number zero is sent to clear all the LEDs.

Each time data is sent to the port, the `WaitForReply`
subroutine is called to ensure that synchronization between
the PC and the BS2 is maintained. Let's see how
`WaitForReply` works.

A repeat-until loop is used to wait until the input buffer
has received one reply byte. The number of bytes received
can be determined using the command
`CheckSerBuffer`. Once a byte is in the buffer `SerIn` is
used to remove the byte from the buffer. We could check
its value, but for this example, we just ignore it as the real
purpose of the data was synchronization.

The BS2 Program
Of course, the BS2 must also be programmed to handle its
side of the work. We will start by dealing with this simple

output example, and then expand the program so that the BS2 can provide both input and output functionality. The program in Figure 3.3 handles the BS2's end of the communication.

```
'  {$STAMP BS2}
'  {$PBASIC 2.5}
'========== assign comm pins ================
   ReceivePin PIN 0
   SendPin PIN 1
'===== Declare necessary variable ============
   cData VAR Byte
'============================================
Main:
   OUTPUT 8
   OUTPUT 9
   OUTPUT 10
   OUTPUT 11
   DO
      SERIN ReceivePin,84,[cData]
      outC = cData   // store all 4 bits at once
      SEROUT SendPin,84, [cData]
   LOOP
```
Figure 3.3: This program lets a BS2 provide output operations for RobotBASIC

The program in Figure 3.3 is written in PBASIC, a language developed by Parallax for many of their micro controllers. It is not the purpose of this book to teach how to use the BS2; refer to www.Parallax.com for many tutorials and examples for how to program the BS2.

The program starts with comments indicating that the program was written for a BS2 based controller and used version 2.5 of the language. Next, I/O pins 0 and 1 were assigned for receiving and sending the serial data communications with RB. One variable, cData, will be used for the communication.

The Main program starts by configuring pins 8 through 11 as outputs. These pins were chosen because PBASIC already has a group name for these pins (outC). An endless loop then repeats the primary actions of the program.

Inside the loop a SERIN statement waits for a byte on the receive pin. The number 84 in this statement indicates that the communication data rate should be 9600 baud, which is the default rate for RB's serial commands. This translates loosely to a maximum transfer rate of about 960 bytes per second.

Once the data is received, it is copied directly to the four output pins via outC. The data received is then sent back to RobotBASIC to maintain synchronization.

Connecting the Hardware

Current PC's usually do not have standard serial ports, but many manufactures, such as Parallax, provide USB-To-Serial converter. Figure 3.4 shows the Parallax USB2SER Development Tool (part #28024). The data signals for this device are standard TTL signals (0-5 volts) making it easy to interface with a microcontroller. (Standard serial ports use -12 and +12 volt signals that have to be conditioned for use with most microcontrollers.

Figure 3.4: The USB2SER tool from Parallax provides a serial port for any PC with a USB port.

A standard USB cable connects the USB2SER device to a USB port on your PC. The opposite end provides four connections of which we will use three (pin functions are labeled on the reverse side for easy identification). The pins we will use are labeled RX, TX, and Vss. Vss is the ground connection, and RX and TX are the receive and transmit pins. You should connect the RX pin to I/O pin 1 on the BS2, thus connecting the BS2 transmit pin to the receive pin on the serial port. Likewise connect the TX pin to I/O pin 0 on the BS2. The Vss pin should be connected to the Vss pin, the ground connection on the BS2. You should also connect four LEDs to the BS2's I/O pins 8 through 11 similar to the connections made in Figure 2.1 from the last chapter.

The program in Figure 3.3 should be downloaded to the BS2 (refer to Parallax documentation) and the program in Figure 3.2 should be running on the PC. If you have done everything correctly, the four LEDs on the BOE should blink at a one hertz rate. If you change the data being sent (currently 15) to other numbers from 1 to 14 you can make different LEDs blink.

Adding an Input Port

Now that we have an output port working, and you understand some of the principles of serial communication, let's modify the BS2 program so that it can perform both output and input functions for us. The modified program is shown in Figure 3.5. There are several things to notice.

First, a second variable (rData) has been added to be used for the return data. Next, four I/O pins have been set as INPUT pins. This is not really necessary, as the I/O pins default to inputs, but it makes the intent clear.

When the number is obtained from the serial connection, it is checked to see if it is less than or equal to 15. Let's see why. Since we are implementing a 4-bit output port, the data sent should have a value from 0 to 15. If the BS2 receives data in this range it will assume an output operation is being requested and thus sends that data to the output pins just as it did in the previous program. It also copies the data received into rData, but more on that in a moment.

If the data received from RB is greater than 15, the BS2 program assumes that an INPUT operation is being requested.

When this happens the data is copied from the four input pins to rData (using IND, a label PBASIC uses to refer to all four of these pins). At this point, the value of rData is returned to RB. Notice that if an output operation is performed, the data returned is the same as the data received, but if an input operation is requested, the data from the input pins is sent back.

```
' {$STAMP BS2}
' {$PBASIC 2.5}
  ReceivePin PIN 0
  SendPin PIN 1
  cData VAR Byte
  rData VAR Byte
Main:
  OUTPUT 8
  OUTPUT 9
  OUTPUT 10
  OUTPUT 11
  INPUT   12
  INPUT   13
  INPUT   14
  INPUT   15
  DO
     SERIN ReceivePin,84,[cData]'receive command
     IF cData<=15 THEN
        outC = cData
        rData = cData
     ELSE
        rData = IND
     ENDIF
     SEROUT SendPin,84, [rData]
  LOOP
```

Figure 3.5: This code programs a BS2 to act as I/O ports for RB.

Once the controller has been programmed in this way, there is no need to program it again, unless you want to change its functionality. As long as you want the BS2 to act as a 4-bit input and a 4-bit output port, no further programming is required for the BS2. Let's see how RB can use these new ports.

Creating New I/O Functions
In order to make it easy to utilize the new ports, we will create two special functions to handle the details for us. A demo program that includes the two new functions is shown in Figure 3.6.

```
Main:
  setcommport 33
  xyText 10,10,"Set Output Port Status:",,\
       20,fs_Bold|fs_Underlined
  for i=0 to 3
    addcheckbox ""+i,430+20*(2-i),20," "
  next
  xyText 90,70,"Input Port Status:",,\
           20,fs_Bold|fs_Underlined
  rectangle 405,70,493,108
  while true
    n=0
    for i=0 to 3
      if getcheckbox(""+i) // see if each
                           //box is checked
        n = n+2^i
      endif
    next
    call Soutput(n) //outputs n to the port
    call Sinput(n)  //asks for input and store in n
    // display the data in binary
    n= bin(n,4) // convert to a binary string
    xyText 417,75,n,,20,fs_Bold
  wend
end

sub Sinput (&d)
  SerialOut 16
  repeat
    CheckSerBuffer nb
  until nb=1
  serin d
  d=ascii(d)  // convert from character to a number
return

sub Soutput (Sdata)
  SerialOut Sdata
  repeat
    CheckSerBuffer nb
  until nb=1
  serin  d
  if ascii(d)<>Sdata then print "output error" \end
return
```

Figure 3.6: This program creates and demonstrates two I/O functions
for interfacing with the BS2 I/O ports.

Creating Custom I/O Functions

RobotBASIC has the ability to create standard *gosub*-style subroutines, but it also has the ability to create *function*-style subroutines that utilize local-scoped variables, and can be passed parameters. Let's look first at the serial output routine, Soutput, we created at the end of Figure 3.6.

This new type of subroutine is specified using the sub designator followed by the routine name and then a list of the parameters it requires in parentheses. In this case, the data passed to the routine will be the data intended for the output port. That data will be called Sdata throughout the subroutine.

The routine starts by sending Sdata to the BS2 using the serial port. The routine then waits for a byte to come back, and compares that byte to the data sent. If it does not match, then an error message is printed and the program is terminated.

The serial input function, Sinput, is different right from the start, as the parameter name used to refer to the passed data is proceeded by an &. This tells RB that any references to this variable should actually reference the original variable used to call the routine. This will become clearer shortly. For now let's just see how the routine works.

It starts by sending the number 16 to the BS2 using the serial port. Remember that we programmed the BS2 so that when it gets numbers greater than 15 it assumes that an input operation has been requested, and sends the port data back over the serial connection. The Sinput routine waits for this data and places its ASCII value into the variable d, which really means it is placed into the variable specified when Sinput was called. Again, more on this shortly.

Now that the two routines have been examined, let's see how they are actually used in the main program. In this example the program is very similar to the one in Figure 2.10, so very little explanation is necessary.

The only new aspects of the program are the calls to the new functions. When the program needs to output the data n, for example it used this line.

```
call Soutput(n)
```

A call-statement is used to invoke function-style subroutines. In this example, the value of n is passed to the routine.

The next line in the program requests input data from the BS2 using the following statement.

```
call Sinput(n)
```

Again the variable n is specified as an argument, but any name could have been used. Recall that the `Sinput` routine preceded the variable name (d) of the data passed with a &. This means that when data is placed into the variable d inside the `Sinput` routine, it is really being placed into the variable n.

> ⓘThe use of the & operator is to designate the parameter as a *by-reference* parameter as opposed to a *by-value* parameter. When a caller statement passes a variable in the position of a by-reference parameter then the value of that variable can be changed within the subroutine. This is useful for passing data back to the caller as we did above.

The lines following the call to `Sinput` just work with the variable n, because it will hold the value of the data gathered by the BS2. This functionality is called *passing-by-reference* because the subroutine will reference the original variable whenever the designated variable is used.

Conclusion
At this point you have numerous options for sending data to, and gathering data from, external devices. In the

examples so far, we have only controlled the state of LED's and obtained the status of switches. In the next two chapters we will learn how to control motors and how to gather sensory information.

Chapter 4

Motor Control

Now that you have some understanding of how ports work, it is time to start doing more interesting things. In this chapter we will see how output ports can be used to control motors. In particular, we will examine both DC motors and servomotors, as these are the most commonly used motors for hobby-oriented projects.

DC Motors

A typical DC motor runs at a relatively high speed and has minimal torque. For most applications, the motor's usability will be improved by using a series of gears to reduce the speed and increase the torque. Finding appropriate gears and mounting them properly can be time consuming. Fortunately, DC motors with integrated gears can be purchased from many manufactures. Motors with integrated gears are often referred to as *gear-head* motors.

Lynxmotion.com sells a variety of motors similar to the one in Figure 4.1. Other gear-head motors can be obtained from Pololu.com and Parallax.com. In many cases, wheels that mate with the motors are also available. The motors come in many sizes and current requirements. Small motors might only use 100ma of current, but larger motors can easily require many amps. Unfortunately, most port pins are capable of delivering only a few milliamps of current.

Figure 4.1: Gear-head motors provide excellent torque.

Buffering the Port

In order to control a motor from a port pin, we need some form of buffering circuitry. Figure 4.2 shows how an NPN transistor can be used to increase the current capability of a port pin. Whenever the port outputs a 1, the transistor will turn on and conduct current to the motor (or other devices such as a solenoid or a light bulb). Unfortunately, sending a 1 or 0 only allows the motor to be turned on and off. This simple action cannot control the speed of the motor, nor can it control the motor's direction.

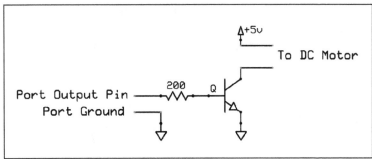

Figure 4.2: This transistor circuit allows a port pin to control a high-current device.

Controlling Speed

The best way to control the speed of a DC motor is to turn it on and off in a predetermined manner. Let's look at the waveforms below and assume it is being generated by the program fragment in Figure 4.3.

```
// this fragment turns the motor on 25% of the time
while true
    call Soutput(1)  // turn motor on
    delay 1
    call Soutput(0)  // turn motor off
    delay 3
wend
```

Figure 4.3: This sample code generates a
waveform with an ON time of 25%.

In general, if such a waveform was applied to the motor, the motor would run at about 25% of its normal speed. You could change the speed by altering the delays to increase or decrease the percentage of on time. Ideally, we would want *at least* 50 pulses per second so that the motor will run smoothly. Often a higher frequency is used.

Controlling the Direction

In order to control the direction of the current through the motor, four transistors must be connected in what is called an H-bridge. For such an arrangement, one output pin can control the speed, as described above, and another can be used to specify the direction (for example, 0 for forward, 1 for reverse).

The building of an H-bridge is often difficult for many beginning hobbyists, but fortunately we have a solution. Many manufactures provide motor controllers that not only contain appropriate H-bridge components, but an integrated micro controller as well. This makes it very easy for RobotBASIC to control the speed of a motor.

Figure 4.4 shows the Robotclaw 2x5A motor controller from Lynxmotion. It handles two motors of up to 2.5 amps each. This controller has many features far beyond the scope of this introductory text, so refer to the online documentation from Lynxmotion for more information.

Figure 4.4: This Lynxmotion motor controller can handle two motors.

One nice thing about the controller pictured in Figure 4.4 is that it has a Simple Serial mode that makes it easy for even beginners to control motors from a RobotBASIC program. For this example, it is assumed that you have connected the controller to your PC with a serial interface. It is also assumed that motors have been properly attached and the controller configured for Simple Serial Control at 9600 baud (as described in the Lynxmotion documentation).

Before we continue, it is worth mentioning that there are many controllers that can be controlled in a similar manner to our example. Every product has its own commands and features though, so always refer to the accompanying documentation.

In the Simple Serial mode, the Robotclaw 2x5A controller only needs to be sent a single byte, and we already know how to do that with RobotBASIC's SerialOut and SerOut commands. This single byte is capable of controlling the speed of either motor.

Remember, a byte is made up of 8 bits. In this case, the MSBit (most significant bit) determines which of the two motors is being affected by this command. The bottom 7 bits will control the speed and the direction of the selected motor. Since a 7 bit number has a decimal range from 0 to 127, it is easy to imagine that number as the controlling parameter.

If we divide the 7 bit number in half, we get 64 and 64 will be used to STOP the motors. Since the value of the MSBit is 128, Sending 64 alone will stop one motor (Motor 1), and sending 64+128 will stop the other (Motor 2). Numbers smaller than 64 will turn the selected motor in reverse, with each smaller number increasing the speed. Numbers above 64 will turn the selected motor forward, with each larger number increasing the speed. The diagram below makes this clearer.

Data Sent	Action for Selected Motor
127	Full Speed forward
107	Medium forward
85	Slow Forward
64	Stop
43	Slow Reverse
21	Medium Reverse
1	Full Speed Reverse

Each number represents the action for Motor 1. If 128 is added to the number in the table, the action applies to Motor 2. Numbers in between the numbers in the table will create proportional speeds. For example, the number 10 would create a medium-fast speed in reverse, while the number 90 would create a medium-slow forward speed.

To be fully complete, there are two control numbers that do not behave exactly like you might expect. The number zero alone will turn off BOTH motors simultaneously and the number zero with the high bit set (making it 128) will turn Motor 2 on in full reverse.

Now that you see what numbers have to be sent to create certain motor actions, let's look at an RB program for generating those numbers. Figure 4.5 shows the program. It has been written so that it can be run even if you do not have a serial port currently connected.

```
spNumber = 0 // change to the number of your port
SetCommPort spNumber
AddSlider "Motor 1",100,100,250,1,127
AddSlider "Motor 2",420,100,250,1,127
xyString 200,80,"MOTOR 1"
xyString 520,80,"MOTOR 2"
xyString 100,150,"REVERSE      OFF         FORWARD"
xyString 420,150,"REVERSE      OFF         FORWARD"
rectangle 210,195,260,220,Black,White
rectangle 530,195,580,220

while true
  m1=GetSliderPos("Motor 1")
  xyString 220,200,m1," "
  m2=GetSliderPos("Motor 2")
  xyString 540,200,m2," "
  if spNumber>0
    SerialOut m1
    SerialOut m2+128
  endif
wend
```

Figure 4.5: This program uses sliders to control two motors.

The program in Figure 4.5 creates the screen shown in Figure 4.6. The first two lines set up the serial port. Next, two sliders are installed with a range of 1 to 127 and then boxes are drawn to show the current output from each slider.

An endless loop reads both sliders and displays their current position in the boxes. If a serial port number has been set (by the first line of the program) then the slider numbers are sent to the port. Notice that when the value for Motor 2 is sent, that 128 is added to the value chosen by the slider.

Remember, now that you know how to communicate with external devices using a serial port, you should be able to choose your own motor controller. Just read the documentation to make sure it meets your needs and then send the proper data over the serial link.

It is also worth mentioning that many motor controllers, including the Lynxmotion model discussed above, have a mode that allows them to be controlled as if they are a servomotor, so let's see how to control a servo motor.

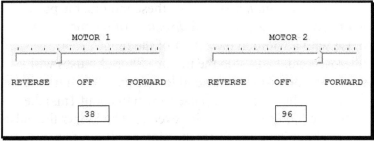

Figure 4.6: Moving the sliders with the mouse will alter the speed and direction of motors connected to the motor controller.

ⓘIf you want to control a stepper motor then have a look at the PDF: http://www.robotbasic.org/resources/RobotBASIC_US Bmicro_U4x1.pdf. In it there is a project similar to the above but using the functionality in the U421 from USBmicro to control a stepper motor even over the Internet (see Chapter 8).

Servomotors
Servomotors are basically a DC motor with an integrated gear train - but they also contain an integrated motor controller. To control a servomotor, you must send it a continuous series of pulses, typically at 50 Hz, but the actual frequency is not critical. What is important is the width of the pulses being sent.

Servomotors come in two basic varieties. Standard servos only turn about 180°. The width of the control pulses determines the position of the servo's output shaft. Typically, a pulse width of 1500 microseconds will center the output shaft. Pulses ranging from 500 to 2500 microseconds generally move the shaft through its full range (the exact range can vary for different servomotors).

Since the pulses sent to a standard servo move it to specific *positions*, they are ideal for building a robot arm.

You can also purchase servomotors labeled as *continuous rotation* servos. For these motors, the pulse width controls the *speed* and *direction* of the motor. A pulse with a width in the center of the range (1500 microseconds) will stop the motor. Shorter pulses will move the motor in reverse and longer pulses will turn it forward. The shorter the pulse (down to about 1ms) the faster the motor will move in reverse. The longer the pulse (up to about 2ms) the faster the motor will run forward.

Unfortunately, RobotBASIC is not capable of creating accurate microsecond pulses so help is needed to control servomotors.

Microcontrollers Can Help

We could certainly program a microcontroller to create servo pulses based on numbers that we send it, thus allowing serial commands to control a servomotor. Fortunately, companies have already created pre-programmed controllers that can control servomotors via serial commands.

Micro Maestro 6-Channel USB Servo Controller

The Micro Maestro 6-Channel servomotor controller from Pololu is shown in Figure 4.7. It can be interfaced via the 5-volt serial port discussed in Chapter 3, but more importantly, the Micro Maestro also has a direct USB interface making it very easy to use with RobotBASIC.

In the USB mode, it actually creates two separate virtual serial interfaces, a COMMAND port and a TTL port. You can find the numbers for both of these ports using the methods discussed in Appendix B. The command port allows more advanced operations, but the TTL port is all we need to control this device, especially for this introductory text.

Figure 4.7: This servomotor controller from Pololu can be interfaced via a USB port.

Once you have the port number for the TTL port, you can control six servomotors using RobotBASIC's `SerialOut` command. To position a motor, you simply tell the controller the width of the pulses you want sent to that motor. The controller handles all the details. It sends each servomotor the specified width pulse, and it will continue to do that for you at the appropriate frequency, all by itself. You simply send it a new pulse width when required.

The Micro Maestro has several control modes, but the simplest is called the Compact Protocol. The Compact Protocol does not have any frills, but it provides all the control many people need. Using this simplified mode, RB needs to send four bytes to control any one of the six servos. These four bytes take the following form.

The first byte is always decimal 132 or 0x84. This byte is a command code that tells the controller what you are trying to do. The second byte in the sequence is the number of the servo you wish to control. For the Micro Maestro this would be from a number from 0 to 5.

The next two bytes in the sequence specify the desired pulse width for the indicated servo. In order to provide a high degree of accuracy, the number sent is in units of ¼ microsecond. This means that if you wish to specify a pulse width of 1500 microseconds, you will have to send the number 6000 or 1500*4.

Since the largest number you can store in an 8-bit byte is 256, we will need two bytes to hold the entire number. The number 6000 (decimal) can be represented as this 16-bit binary number.

00010111 01110000

If we break that into two 8-bit bytes, we get these two binary numbers and their decimal equivalent.

$$0\%00010111 = 23$$
$$0\%01110000 = 112$$

Since it is required that we send the LSByte first, we would send a byte valued at 112 followed by a byte valued at 23. This could be accomplished with the following command.

```
SerialOut 132,0,112,23
```

Before you start thinking that converting the numbers is a lot of work, remember we are using a computer. We can let it do the work. If the variable **n** contains the number of quarter microseconds for servo 0, we can use the following statements to accomplish our goal:

```
SerialOut 132,0, n, n >> 8
```

Putting it all Together

The program in Figure 4.8 should be easy to follow. The data statement in the program is an easy way for RB to create an array and initialize its elements.

Let's assume that the program is manipulating a robot arm where each joint is controlled by one of the six servomotors. Just insert your code that decides what each joint of the arm is supposed to do. For example, your code could read sensors on the arm to decide on actions to take, or it could use sliders so the arm could be controlled with the mouse. Once the position of joints has been entered into the array (servos that do not have to move do not have

to be changed) then the `for`-loop sends the desired servo
positions to the controller and it does the rest. Of course, it
is only necessary to send out changes for the servo
positions, but the `for`-loop just updates everything.

```
SetCommPort 0 // change to actual port number
data n; 6000,6000,6000,6000,6000,6000
while true
   // your program can decide HERE where it wants each
   // servo to be and change the proper values in
   //the array
   for i=0 to 5
      SerialOut 132,0,n[i],n[i]>> 8
   next
wend
```

Figure 4.8: This program template shows how RobotBASIC can
control six servomotors through an external servo controller.

Summary

In this chapter you have seen how RobotBASIC can control
both DC and servomotors. Remember, we have only
touched on the fundamental principles to get you started.
Now that you understand how the serial port communicates
with external devices you can interface RB with many
products from many manufacturers. Refer to their
documentation and learn to use the advanced features of
your chosen product to enhance your projects.

Chapter 5

Sensors

Previous chapters provided information on how data can be obtained from external devices using both serial and parallel ports. The purpose of this chapter is to introduce you to various types of sensory devices. Once you grasp the principles of using them, you should be able to utilize a wide variety of products from various manufactures by reading their specifications sheets.

Modern Sensor Technology
In the recent past, if you wanted some type of sensor, you had to build it yourself. This often required knowledge of operational amplifiers, filters, phase lock loops, as well as many other aspects of electronics. Nowadays, though, manufacturers offer a wide variety of sensors that can be interfaced easily to any I/O port with only a minimal knowledge of electronics.

Types of Sensors
The types of sensors available today seem almost limitless and there is not enough room in this introductory text to cover them all. RB's simulated robot has numerous sensors that provide a reasonable representation of what is available. Therefore, this chapter will focus on how to implement many of the sensors available on the simulator.

Bumper Sensors

Bumpers are generally used to give a robot the ability to determine when it makes contact with obstacles around its perimeter. The simulator only uses four bumpers, because they are viewed as a fail-safe sensors. If everything is working properly, a well behaved robot should not actually collide with objects. If something should make its way past other sensors though, the bumpers should provide a dependable detection system.

Bumpers can be implemented using nothing more than switches, interfaced as we saw in Figure 2.4. Any type of switch can be used, but snap action switches with a long arm, such as the one in Figure 5.1 from Pololu, make creating a bumper easy.

Figure 5.1: Switches with a long lever arm can be used as tactile sensors such as bumpers.

IR Perimeter Sensing

As stated earlier, ideally, a robot should never actually collide with obstacles. One way of accomplishing this is to use an Infrared LED to project IR light away from the robot. If an object is near, some of the light will be reflected back towards the robot. A phototransistor can be used to detect the light and create a logical 1 or 0 that can be read using an input port.

In the old days the IR sensors built in this way by hobbyists were often unreliable because bright sunlight, or even florescent lights could confuse the sensor because both contain high amounts of IR frequencies. This problem can be solved by modulating the infrared light being emitted with a lower frequency and then using appropriate filters to restrict the phototransistor from detecting other frequencies. As mentioned earlier, constructing such a sensor would require knowledge of electronics.

Fortunately relatively low-cost IR sensors with all these features are now readily available. Figure 5.2 shows one from Pololu. Actually they offer two varieties of this sensor, one that detects objects up to 2 inches and another that detects up to four inches. It is important to know that these sensors do NOT tell you how far an object is from the sensor, only if an object is within the sensor's range.

Figure 5.2: Pololu offers two varieties of
IR proximity sensors.

Not only does this sensor contain all the electronics to make it reliable, it is extremely easy to use. It has only three pins and two of them are used for 5-volt power and ground. The third pin is the output signal which can be connected directly to an input pin on any of the input-port types discussed in earlier chapters. The pin can be read just like a switch input: a 0 means that an object is within range and a 1 means no object is detected.

RobotBASIC's simulated robot has the equivalent of five of these sensors spaced equally across the front half of the robot. Our book, *Enhancing the Pololu 3pi with RobotBASIC* shows how to modify a standard 3pi so that it has most of the sensors found on the simulator. It is an advanced project, but something you might want to consider after you have mastered the principles discussed in this book. A picture of that robot showing both the IR sensors and bumper sensors made from switches is shown in Figure 5.3.

Figure 5.3: This modified 3pi robot has nearly all the sensors of the RobotBASIC simulator.

Ranging Sensors

Sensors that can measure the distance to objects (rather than just detect if an object is present) can be useful. One way to do this is with the Ping))) ultrasonic sensor from Parallax shown in Figure 5.4.

Figure 5.4: The Ping))) Sensor from Parallax measures the distance to objects.

The sensor works by sending out a high frequency sound wave through one of its transducers (much like a speaker on your stereo). It then waits for the sound to be reflected off an object (the sound is detected with another transducer acting as a microphone). Circuitry in the sensor puts out a pulse that stays high until the sonar comes back or it times out. This pulse duration can be measured using a suitable microcontroller such as the BS2. The pulse duration is proportional to the distance the object is from the sensor.

One of the great things about Parallax is that they provide massive amounts of documentation and programming examples with their products. For example, they provide a sample program that gathers data from the Ping))) sensor. It is easy to modify their program so that it sends the data to RobotBASIC over a serial pin, just as we did in Chapter 3 when we use the BS2 to create parallel I/O ports. The modified program is shown in Figure 5.5. Compare it to the program in the zip file titled BASIC STAMP EXAMPLE CODE found on the Ping Sensor product page at Parallax.com. Study how our new program was derived from the original. Learning how to use the sample code Parallax provides for all its products can help you utilize other Parallax products just as we have in this example. Let's look at Figure 5.5.

Most of the work is done by the subroutine Get_Sonar which was taken almost verbatim from Parallax. If you read the Parallax documentation for the Ping Sensor, you can understand how this subroutine works, but the nice thing is that you do not have to understand it. Parallax offers many sensors far more complex than the Ping, and the code they provide to read the sensors is often complex. As long as you can identify the section of code that performs the actions you want though, you can paste that code into your own programs and accomplish your goals-without having an in depth understanding of the technical details.

```
'    {$STAMP BS2}
'    {$PBASIC 2.5}

Ping        PIN     15
Trigger     CON     5        ' trigger pulse = 10 uS
Scale       CON     $200     ' raw x 2.00 = uS
RawToIn     CON     889      ' 1 / 73.746 (with **)
IsHigh      CON     1        ' for PULSOUT
IsLow       CON     0
rawDist     VAR     Word     ' raw measurement
inches      VAR     Word
cData       VAR     Byte

Main:
  DO
    SERIN ReceivePin,84,[cData]
    GOSUB Get_Sonar            ' get sensor value
    inches = rawDist ** RawToIn ' convert to inches
    IF inches>255 THEN
      cData=255
    ELSE
      cData = inches//256  'keep lower 8 bits
    ENDIF
    SEROUT SendPin,84, [cData]
  LOOP
END

Get_Sonar:
  Ping = IsLow                ' make trigger 0-1-0
  PULSOUT Ping, Trigger       ' activate sensor
  PULSIN Ping, IsHigh, rawDist ' measure echo pulse
  rawDist = rawDist */ Scale  ' convert to uS
  rawDist = rawDist / 2       ' remove return trip
RETURN
```

Figure 5.5: This program allows RobotBASIC to interface with a Parallax Ping sensor.

The Main code in Figure 5.5 waits for a byte from RobotBASIC before calling Get_Sonar to obtain the time reading from the Ping sensor. Then the time is converted to inches just as the original Parallax code did. It is worth mentioning that the Parallax code appears far more complex because it was written to work with several different Parallax controllers. Our code is less complex because it *only* works with a standard BS2.

Parallax's code stores the final answer in the variable `inches`, which is a 16-bit word. Since the maximum range of the Ping))) is about 120 inches then the answer should fit in an 8-bit word, so our code uses an IF-block to transfer the answer to the byte variable `cData`. If for any reason the original distance was greater than 255, then it is set to 255.

Finally, the program sends the 8-bit data back to RB via the serial port. Study this code to make sure you understand its operation. If some things do not make sense, you may have to read Parallax's programming manual for the BS2 (available for download on their web site).

Building a System

Up till now, all of our examples have been simple stand-alone applications. In a realistic project, you will need to combine many of the principles in this text to create a system for accomplishing your goals.

Remember the BS2 program from Chapter 3 (see Figure 3.5) that created a 4-bit input and a 4-bit output port that could be accessed from RB. Imagine combining that program with the Ping program from this chapter, providing a system allowing you to read the Ping Sensor as well as providing the ability to use the I/O ports for other purposes. The combined program is shown in Figure 5.6.

Study the program carefully to see how the original two programs were combined. Also notice how the program uses the byte sent to the BS2 to decide what to do. As before, if the number sent is less than 16, it is assumed to be data for the output port. Now, if the data is 16, it assumes RB is requesting data from the input port. And, if the data is 17, it is assumed that RB wants a distance measurement from the Ping sensor.

```
' {$STAMP BS2}
' {$PBASIC 2.5}
ReceivePin  PIN 0
SendPin     PIN 1
Ping        PIN 2
Trigger     CON    5       ' trigger pulse = 10 us
Scale       CON    $200    ' raw x 2.00 = uS
RawToIn     CON    889     ' 1 / 73.746 (with **)
IsHigh      CON    1       ' for PULSOUT
IsLow       CON    0
cData       VAR    Byte
rData       VAR    Byte
rawDist     VAR    Word    ' raw measurement
inches      VAR    Word

Main:
  OUTPUT 8
  OUTPUT 9
  OUTPUT 10
  OUTPUT 11
  INPUT  12
  INPUT  13
  INPUT  14
  INPUT  15
  DO
    SERIN ReceivePin,84,[cData] 'receive command
    IF cData<=15 THEN
       OUTC = cData
       SEROUT SendPin,84, [cData]
    ELSEIF cDATA=16 THEN
       rData = IND
       SEROUT SendPin,84, [rData]
    ELSEIF cData=17 THEN
       GOSUB Get_Sonar          ' get sensor value
       inches = rawDist ** RawToIn ' convert
                                   ' to inches
       IF inches>255 THEN
         rData=255
       ELSE
         rData = inches//256  'keep lower 8 bits
       ENDIF
       SEROUT SendPin,84, [rData]
    ENDIF
  LOOP
END

Get_Sonar:
  Ping = IsLow                 ' make trigger 0-1-0
  PULSOUT Ping, Trigger        ' activate sensor
  PULSIN Ping, IsHigh, rawDist ' measure echo pulse
  rawDist = rawDist */ Scale   ' convert to uS
  rawDist = rawDist / 2        ' remove return trip
RETURN
```

Figure 5.6: This PBASIC program lets RobotBASIC access I/O ports
as well as the Ping sensor data.

RobotBASIC programs can communicate with this program just as we did in Figure 3.6, except now, when 17 is sent to the BS2, the Ping))) data is returned.

The principles demonstrated by the program in Figure 5.6 should not be taken lightly. Remember, the BS2 has 16 I/O lines and thus far we have only used 11 of them. This means we have 5 more lines for adding more Ping sensors or other sensors.

Line Sensors

The RobotBASIC simulated robot has sensors beneath it that can sense when they are over a line. The QTR-1RC from Pololu (see Figure 5.7) can provide this capability for a real robot. This is an individual sensor, but Pololu also offers it in an eight-sensor array on a single board. Pololu also offers these same sensors in analog versions, but only the digital version is suitable for our use.

The QTR-1RC sensor is very easy to use. It has only three pins, +5volts, ground, and the output signal. Once power is connected (make sure the sensor's ground is connected to the I/O port or controller's ground), simply connect the sensor's output pin to an appropriate pin on whatever input pin you are using. When the pin reads a 0, it means the sensor is over a reflective surface (such as a white poster board). A 1 means a non-reflective surface, such as black electrical tape, has been detected.

Figure 5.7: This Pololu sensor can detect a line beneath a robot.

An Electronic Compass

Parallax offers two electronic compasses. The HMC6352 shown in Figure 5.8 is best suited for our needs. It is accurate to less than one degree, and may be interfaced using I^2C communication.

Figure 5.8: The HMC6352 electronic compass.

As expected, Parallax provides routines for interfacing the HMC6352 with a BS2. If you understood the principles used to interface the Ping))) sensor, you should be able to use the same techniques to interface the compass. Once you get a basic program working, you should consider modifying the program in Figure 5.6 so that it allows RobotBASIC to retrieve the compass heading.

Before you start, there is something that needs to be mentioned. Remember, the program, as written, always returns an 8-bit byte limiting the number returned to a range of 0-255. The compass however, generates a number from 0-359. One easy solution is to use two new codes for retrieving data. Use the code 18 to get the high byte and the code 19 to get the low byte of the two-byte answer. An alternative solution would be to divide the answer by 2 before it is returned to RB (making it fit in a single byte). RB would have to multiply the answer by 2 in order to recreate the data. These operations will reduce the resolution to 2°, but if that is acceptable this is an easy solution. Another solution is to modify the communications mechanism and have the BS2 return two bytes and RB would expect two bytes from the BS2.

Don't assume that you have to use a BS2 for your micro controller. RB can be interfaced with nearly any processor you choose. The principles used here are the same, but you will have to learn the language and idiosyncrasies of your chosen processor. This takes time of course, but your efforts can be well rewarded.

RobotBASIC offers other books that detail how to interface with Parallax's Propeller processor and Pololu's 3pi robot which uses an ATmega328 processor. If you don't need an entire book of instruction, you will also find numerous downloadable PDF tutorials on our website showing further examples for how to communicate in many different ways with the BS2 and the Propeller Chip.

Other Sensors
RobotBASIC's simulated robot has a few other sensors, but the ones covered in this chapter are appropriate for an introductory text. As you work though these examples, you can develop an understanding of the principles needed to utilize a wide variety of sensors by simply studying the documentation available from the manufacturer.

Chapter 6

Speech, the Human Interface

One of the things that separate humans from other animals is our ability to communicate verbally. Robots that can mimic this behavior, even if they are not perfect at it, are usually perceived to possess some level of intelligence.

The Windows operating system already has built-in speech capabilities. The goal of this chapter is to demonstrate how we can utilize these capabilities. We will start with generating speech

RobotBASIC has the ability to play wav files, so we could just record phrases that we wish the computer to speak, and play them back when needed. Such an approach is very limited though so let's look at a better method that can make RB say anything we want.

Synthesizing Speech

Ideally, we would like the computer to be able to say any string. Fortunately there are many low-cost programs (and even a few free ones) that tell Windows to *say* anything copied to the clipboard. One free program is available from:

> http://www.portalgroove.com/speechplayer/

After you install this program, Windows will analyze anything copied to the clipboard and use the rules of pronunciation to convert the text to speech played over the

---

computer's speakers. Since RobotBASIC has commands to copy text to the clipboard, we can easily give our programs the power of speech. Look at the program in Figure 6.1.

You may be surprised at how simple this program is. If you run it (assuming you have the Portalgroove Speechplayer active) you will hear your computer say the phrases specified in the program. The delay in the program is necessary because new phrases will not be recognized if they are received while another phrase is being spoken.

```
SetCBtext("This is a test")
delay 3500
SetCBtext("and So is this!")
end
```

Figure 6.1: RobotBASIC's ability to copy text to the clipboard makes speech synthesis easy.

Voice Recognition

Recognizing the spoken word is almost as easy. The first thing you need to do is enable Windows' Voice Recognition capabilities. This is easily done from the Windows' Control Panel by just following directions. If you have not set up Voice Recognition before on your PC, you should go through the tutorial and training. It takes less time than you might imagine.

Microsoft implies that Voice Recognition can only be used with voice-compatible programs, but that is not entirely true. The RB editor cannot be operated with voice commands, and using the RB HELP file while voice is activated can sometimes cause RB to stop responding, but it is easy to use speech input with RobotBASIC applications.

Let's look at a simple example to see just how easy it can be. The program in Figure 6.2 will let you enter data from the keyboard and then print whatever you entered. If you have Voice Recognition activated, then this same program will print whatever you say (RB's terminal screen must be the active window). The only drawback with this program is that you *must* say ENTER after you have voiced your data. Saying ENTER is equivalent to pressing the ENTER key, which is necessary for the INPUT statement. We will eliminate the need to say ENTER shortly.

```
while true
   input a
   print a
wend
```

Figure 6.2: This program will work with the
keyboard or with voice input.

You can easily add voice output to the program in Figure 6.2. Instead of printing the variable a, just copy it to the clipboard. The program will then say everything that you say.

This program can help you see how well (or how poorly) your speech is recognized. If you say "raise your arm", for example, the system might think you said "raison on". Speak distinctly using a quality microphone and keep it at the same distance from your mouth for the best results.

Let's see how we can use the voice data to alter how our program operates. The example program in Figure 6.3 will also demonstrate how to handle words that Windows' does not recognize properly.

There are several things that are important here. First notice that we are comparing the input phrase to words that have their first letter capitalized. This is necessary because the voice recognition software automatically capitalizes the first word in each sentence. We will see better ways of handling this shortly, but addressing these points can help you see how to handle voice recognition in your own programs.

```
while true
  input t
  if t="Blue"  then clearscr blue
  if t="Read"  then clearscr red
  if t="Green" then clearscr green
  print t
wend
```

Figure 6.3: This program allows your voice to change the color of the screen (if voice recognition has been activated).

The second thing you should notice in the program is that we are checking for "Read" rather than "Red". This was necessary because when you say "Red", Windows' will give you the homonym "Read" (as in "I read the book yesterday"). If you ask Windows' to say the same word it could pronounce it differently again as in "I will read the book tomorrow"). As long as we know this is happening, we can adjust our software to handle it. Before we do that though, let's see how to eliminate the need for saying ENTER.

Eliminating the Need for ENTER

The reason we need to say ENTER is because the INPUT statement waits for additional characters until it receives a carriage return (ASCII code 13), which occurs when you press the ENTER key.

RobotBASIC has the ability to create Text Boxes and monitor those boxes for changes made to the text. Since a carriage return is not necessary to obtain the text from Text Boxes, we can use them to improve our voice input. The program in Figure 6.4 shows how this can be done.

```
a="My Edit Box"
AddEdit  a,100,100,500
FocusEdit a
while true
  if EditChanged (a)
    t=GetEdit(a)
    if t="Blue"  then clearscr blue
    if t="Read"  then clearscr red
    if t="Green" then clearscr green
    print t
    SetEdit a,""
  endif
wend
```

Figure 6.4: This program allows voice control without the need to say ENTER at the end of each phrase

The program in Figure 6.4 starts by creating an Edit Box and forcing focus on it. An endless loop constantly checks if the data in the box has changed. If it has changed, we obtain the data and use it as we did in the previous program. At the end of the loop, the text in the Edit Box is cleared to wait for the new data.

An Example Voice Recognition Application

The program in Figure 6.5 shows how the simulated robot can be controlled with voice commands. There are a number of principles demonstrated in the program that warrant discussion.

The `Initialization` routine sets up the Edit Box as well as variables to be used in the program. Finally, it creates the robot near the center of the screen.

When the data is obtained from the Edit Box, it is converted to all caps so that we do not have to worry if the word being compared is at the beginning of the phrase. Next, three variables, `Action`, `Amount`, and `Again` are initialized. As the voice data is analyzed, these variables will be set to values that reflect the action being requested by the user.

The function `InString()` is used to look through the entire string to see if a particular word or phrase is present. When appropriate phrases are found, the variables mentioned above are set appropriately. Notice this allows many different phrases to easily cause the same action.

Once the control variables are set it is easy to utilize them to actually control the robot. Most of the control actions in the program are easy to follow, but perhaps the *Again* section deserves a comment. It allows the user to say things like "repeat the last action" or "a little more". It uses the variable `LastAction` (which keeps track of the last thing the robot did) to enable the program to add something to the last action. For example, after you tell the robot to "move forward", you can then say something like "not so much" and it will back up a little.

Your robot will seem far more intelligent if you do not have to use very specific phrases to control it. You want it to appear that you can say almost anything and the robot will interpret it correctly.

It will take far more work to make the robot truly intelligent, but the tricks demonstrated in this program can dramatically improve how intelligent your robot will *seem*.

There are some powerful concepts demonstrated in this program, so study it carefully. It can give the illusion that totally free forms of speech can be used to control the robot. Let's look at some examples.

You can make the robot move forward using many phrases. Here are a few examples.

forward
move forward
go forward
move ahead

You can also add qualifiers like this.

move forward a little
go forward a lot

Similar actions can be used to make the robot turn.

turn Left
right
left a little
turn right a lot

```
Main:
  gosub Initialization
  while true
    if EditChanged (a)
      t=Upper(GetEdit(a))
      xyString 200,50,t,spaces(27)
      delay 1000
      // analyze the command
      Action=0 \ Amount=35 \ Again=0
      for i=1 to numparts(ActionW,"|")
        w = extract(ActionW,"|",i)
        if Instring(t,w) then Action = Actions[i-1]
      Next
      for i=1 to numparts(AmountW,"|")
        w = extract(AmountW,"|",i)
        if Instring(t,w) then Amount = Amounts[i-1]
      Next
      for i=1 to numparts(AgainW,"|")
        w = extract(AgainW,"|",i)
        if Instring(t,) then Again = Agains[i-1]
      Next
      if Action=Forward   then rForward Amount*2
      if Action=Backward  then rForward -Amount*2
      if Action=LeftTurn  then rTurn    -Amount
      if Action=RightTurn then rTurn     Amount
      if Again
        if LastAction=Forward
          if Again=More then rForward 15
          if Again=Less then rForward -15
          if Again=DoOver then rForward Amount
        endif
        if LastAction=Backward
          if Again=More then rForward -15
          if Again=Less then rForward 15
          if Again=DoOver then rForward -Amount
        endif
        if LastAction=LeftTurn
          if Again=More then rTurn -15
          if Again=Less then rTurn 15
          if Again=DoOver then rTurn -Amount
        endif
        if LastAction=RightTurn
          if Again=More then rTurn 15
          if Again=Less then rTurn -15
          if Again=DoOver then rTurn Amount
        endif
      endif
      if Action and Action<>DoOver then LastAction=Action
      SetEdit a,""
    endif
  wend
End
```

Figure 6.5: This program allows voice control over the simulated robot. The Initialization module is shown in Figure 6.6

```
Initialization:
  a="My Edit Box"
  AddEdit   a,100,20,600
  FocusEdit a
  Forward   = 1
  Backward  = 2
  LeftTurn  = 3
  RightTurn = 4
  DoOver    = 5
  More      = 6
  Less      = 7
  LastAction = 0
  rLocate 400,300
  rSpeed 20
  Line 0,75,800,75
  ActionW  ="FORWARD|FORD|AHEAD|I HAD|"
  ActionW +="BACK|LEFT|RIGHT|WRITE"
  data Actions;Forward,Forward,Forward,Forward
  data Actions;Backward,LeftTurn,RightTurn,RightTurn
  AgainW ="MORE|LESS|LAST|MUCH|AGAIN|REPEAT"
  data Agains;More,Less,Less,Less,DoOver,DoOver
  AmountW = "LITTLE|LOT"
  data Amounts; 20,80
return
```

Figure 6.6: This Initialization module should be included with the Main program in Figure 6.5.

You can cause the last action to be repeated with commands like these.

repeat
repeat the last action
do that again

You can even qualify actions. The following statements will make the robot move forward then back up a little.

move forward a lot
not so much

Similarly, these statements will make the robot turn right then turn a little more.

turn right
a little more

Since the program is only looking for key words inside the phrases, you can even be polite like this.

turn to the right a lot please

Now that you know how easily speech can be used with your programs, you can start making your robot seem even smarter.

Chapter 7

Vision

Giving our computer programs the power of vision is an exciting goal. Such a prospect would have required expensive equipment and professional skills in the recent past.

Nowadays hobbyists have many low cost options for experimenting with vision. This chapter will examine RobotBASIC's no-cost, built-in image processing commands as well as exploring a low-cost commercial option called RoboRealm. Detailed information, as well as a 30-day trial copy of RoboRealm, can be obtained from www.RoboRealm.com.

Inexpensive Web Cams
Low-cost web cams now make it easy to experiment with vision, but just having the hardware is not enough. Your program must be able to capture the image, and then process it. RobotBASIC has commands that can interface with TWAIN-compliant web cams, thus solving the first problem. A commercial program called RoboRealm can interface nearly any camera with RobotBASIC. We will examine RoboRealm later in this chapter.

Analyzing the Image
Let's see what vision processing entails. An image to be analyzed can be thought of as three, two-dimensional

---

arrays (one each, for red, green, and blue information). The numbers stored in each element of these arrays represent the brightness of a respective x-y coordinate in the image. Processing and analyzing the image involves applying mathematical algorithms to these arrays. Let's look at some examples.

An image can be made brighter by increasing all of the array elements by some percentage. A single color pixel in the image can be converted to gray scale by replacing the element in each of the three arrays that is associated with that pixel with an average of those same three elements. These operations are relatively simple but their complexity can increase quickly as more elaborate actions are needed. Converting a standard image to one showing only the edges of objects, for example, involves finding pixels whose properties differ significantly from at least one of its surrounding pixels.

The level of programming and mathematical skills needed to implement sophisticated vision algorithms makes it difficult for many hobbyists to experiment with this subject. Fortunately, RB provides many commands and functions that can help you process and analyze your image. If you need more advanced image processing capabilities then RoboRealm offers even more sophisticated solutions.

RobotBASIC's Vision Capabilities

Let's look first at a vision application implemented entirely in RobotBASIC. The program is shown in Figures 7.1 through 7.5. Some of the lines in the figures were too long to fit in the window, so the \ character was used to indicate that a line is continued. You may enter the data in this way or you may just use long lines. RobotBASIC will handle it either way.

The main program is fairly self-explanatory as the subroutines handle all the details.

```
MainProgram:
  gosub SetColors
  gosub Initialize
  gosub Demo
end
//=======================================================
Demo:
  SetColor Black,BG
  print
  print
  print spaces(28)+"SERVO MOTOR SIMULATION"
  print
  print " This program simulates a servo motor",
  print " which can be moved to specific positions"
  print " by sending it pulses between 1000 and 2000",
  print " microseconds.  You do this by setting the"
  print " variable PulseWidth to the desired number and",
  print " calling the subroutine Servo.  As the"
  print " motor moves, it moves two eyes that will follow",
  print " the object being held.  You will see"
  print " a picture on the screen of what the 'eyes' are",
  print " seeing (it will be like looking in a"
  print " mirror though (when the object is on the right",
  print " side of the picture, it is on the left"
  print " side as viewed by the eyes.  A small square",
  print " will show you where the program thinks"
  print " it is seeing the object.  If it is wrong",
  print " consistently, you need an object with a more"
  print " distinct color."

// now control the servo based on the image
  while true
    call Control(cc,xPosition) // examine the image
                               // and move the motor
    PulseWidth = 2000-(xPosition*100)-100
    gosub Servo
  wend
return
```

Figure 7.1: This is the main portion of the vision demo program.

The subroutine `Control` (Figure 7.2) captures the image from the web cam, and then uses the command `BmpFindClr` to search the image for the color selected when the program is run. In this example the screen is divided into a 10x10 grid. The object's horizontal grid position is calculated and returned to the calling program.

This program uses a picture of a real servomotor and RobotBASIC's 3-D graphic capability to simulate a motor that can position two eyes. Figure 7.4 handles the 3-D transformations needed.

```
sub Control(cc,&x)
    bx=550 \ by=300 \ w=640/3 \ h=480/3
    CaptureImage()
    reSizeBmp "",w,h,"TempFile"
    ReadBMP "TempFile",bx,by
    // see if selected color is within view
    searchClr = cc
    BmpFindClr "",searchClr,Found,SecCnt,.05,.05,10,CntInfo
    // if found, track that one color
    if Found > -1
        // look toward the object based on its position
        x = Found#10 // the horizontal position (0-9)
        y = Found/10 // the vertical position (0-9)
        erectangle bx+(x*w/10),by+(y*h/10),bx+\
            (x*w/10+(w/10)),by+(y*h/10+(h/10)),2,White
    endif
    flip
return


//----------------------------------------------------------
SetColors:
    Print "First, choose camera and set its resolution"
    CaptureSrc() \ CaptureDlg() \ clearscr
    Print "The robot 'eyes' will recognize a selected color"
    Print
    Print "The object should have a unique color compared to the background"
    Print
    Print "Be prepared to hold your object in front of the camera"
    Print "After taking a picture of the object, you will be asked to click it"
    Print "    with the mouse (to identify the color to be searched for)."
    Print
    Print "Press ENTER to start the countdown to take a picture"
    Input "Press ENTER",i
    For i = 3 to 0
        delay 1000 \ clearscr
        XYText 400,300,i,"Times New Roman",50,fs_Bold
    next
    CaptureImage() \ FitCB
    repeat
        ReadMouse x,y,b
    until b ==1
    delay 100 \ ReadPixel x,y,cc \ print cc
return
```

Figure 7.2: This subroutine tracks the object and moves the servo.

```
Initialize:
  x = 70\y=350
  CurPos = 1000
  Incr = DtoR(90.0/100)
  Dim Mot[4,5]
  Data Points;-10,0,0,0,0, 0,10,0,0,0
  Data Points;25,0,0,0,0, 0,-10,0,0,0
  mCopy Points,Mot
  Data View; 50,120,20,100,400,300
  angle=DtoR(90)\gosub Rot
  ge3Dto2DA Mot,View
  SetColor black,gray
  IndColr = 16756912
  FitBMP "ServoMot",299,270,200,200
  ReadPixel 350,300,BG
  ClearScr BG
  FitBMP "ServoMot",299,270,200,200
  Rectangle 299,270,499,295,BG,BG
  Rectangle 299,270,340,380,BG,BG
  Rectangle 465,270,499,400,BG,BG
  LineWidth 2
  Circle 90,390,160,460,black,white
  Circle 190,390,260,460,black,white
  SaveScr 0,250,530,500
  DelTime = 0
  T = Timer()-200
  Flip on
  ClearScr BG
return
```

Figure 7.3: This initialization routine prepares for the animation.

```
Rot:
  for i = 0 to 3
    geRotVz Mot[i,0],Mot[i,1],Mot[i,3] \
              ,angle,a,b,c
    Mot[i,0]=a
    Mot[i,1]=b
    Mot[i,3]=c
  next
return
```

Figure 7.4: This routine uses RB's 3-D graphic routines to transpose the image added to the motor shaft.

The routine in Figure 7.5 is the module that truly acts like the servo motor. Just set the variable PulseWidth to the desired value and call the routine. The motor will move smoothly from its current position to the desired destination, causing the eyes to move appropriately.

```
Servo:
  // PulseWidth must be between
  // 1000 and 2000 microseconds
  if !Within(PulseWidth,1000,2000) then return
  Moves = Round((CurPos-PulseWidth)/10)
  CurPos=PulseWidth
  for k = 0 to Abs(Moves)
    if Moves>0
      angle=Incr
      x=x+.6
    else
      angle=-Incr
      x=x-.6
    endif
    gosub Rot
    ge3Dto2DA Mot,View
    RestoreScr 0,250
    LineWidth 2
    Line Mot[0,3],Mot[0,4],\
          Mot[3,3],Mot[3,4],1,black
    for i = 2 to 0
      LineTo Mot[i,3],Mot[i,4],1,black
    next
    tx=Mot[2,3]\ty=Mot[2,4]
    Circle tx-4,ty-3,tx+4,ty+3,black,IndColr
    FloodFill2 400,300,IndColr
    circle 397,298,403,302,Black,Black
    Line x,y,x+180,y
    LineTo x+180+40,y+20
    LineTo x+40,y+20
    LineTo x,y
    FloodFill2 x+50,y+2,IndColr
    Line x+110,y+20,x+110,y+25
    Line x+110,y+10,x+110,y-35
    LineTo tx,ty-35      // change
    LineTo tx,ty+35
    Line 125,425,x+50,y+50
    LineTo x+50,y+20
    Line x+50,y+10,x+50,y+5
    Line 225,425,x+150,y+50
    LineTo x+150,y+20
    Line x+150,y+10,x+150,y+5
    Circle 90,390,160,460,black,white
    Circle 190,390,260,460,black,white
    eyeX = 215-.7*(85+x)
    Circle eyeX-7,425-7,eyeX+7,425+7,Black,Black
    eyeX = eyeX+100
    Circle eyeX-7,425-7,eyeX+7,425+7,Black,Black
    flip
  next
return
```

Figure 7.5: This routine moves the simulated servomotor.

When the program is run, it starts by asking you to choose the imaging device you wish to use. If you have a TWAIN-compatible camera installed, it should be listed in the dialog box as shown in Figure 7.6.

ⁱRB requires a special DLL to be present in the RB directory when handling web cams. EZTW32.DLL is included in the RB download from our web page.

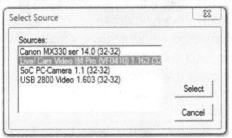

Figure 7.6: RB's vision commands allow you to select the camera you wish to use.

You will also get the chance to preview the images from the camera, allowing you to make any necessary adjustments (see Figure 7.7). At this stage, you want to make sure you get the object to be tracked in the field of view. When you close the dialog shown in Figure 7.7 the program captures an image and asks that you click the mouse on the object to be tracked thus selecting the color that will be used.

Figure 7.7: This dialog allows you to preview the field of view.

When you click the image to be tracked, in this case the red cap on a pen, the program draws the simulated eyes (see Figure 7.8) and moves them to follow the object selected. With most web cams, RB will be able to capture a new image every second or two so don't expect an immediate response to your movements. Even with the slowness, it is interesting to move the object in front of the computer screen and watch the eyes follow it. Since RobotBASIC has access to the image, it is also displayed on the output screen. A box is drawn where the computer thinks the object is. Remember, the image is actually reversed to what the computer sees, so when the object is on the right side of the picture it will be on the left side of your computer (and the eyes should look left).

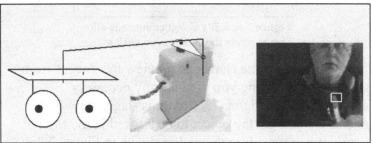

Figure 7.8: They eyes follow objects held in front of the camera.

RoboRealm

As mentioned, the RB commands are not extremely fast, but they do provide an acceptable, no-cost way of experimenting with vision. For more performance you should consider RoboRealm. Even if you want to use RobotBASIC's image processing commands you can capture images much faster with RoboRealm and it can capture images from nearly any camera. Remember, RB capture commands require a TWAIN compliant web cam (you can also use scanners or cameras).

RoboRealm provides a solution for many situations by providing a large library of modules that makes it easy to not only capture images from web cams but process and analyze those images as well.

Although RoboRealm is a programming language itself, it is very different from other languages you might be familiar with. Let's look at some examples to see how RoboRealm can be used to implement a simple vision system. When you first start RoboRealm, you see the opening screen shown in Figure 7.9.

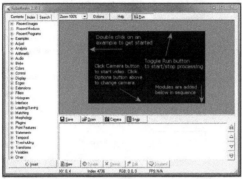

Figure 7.9: RoboRealm's opening screen.

The left side of the screen provides access to the available modules. When a module is selected it is inserted in the programming window at the bottom of the figure. In general, each time an image is acquired from the active camera, all of the selected modules are applied, in consecutive order, to that image. Start by clicking the **Options** button and selecting the camera you are using from a drop down menu. After a camera has been designated, the **Camera** button can then be used to turn it on and off.

In order to demonstrate how easily vision can be implemented let's create a simple program that can track a RED object in the camera's operational view. Expand the **Colors** option in the left column and select **Color_Filter**. When you do you will get the window shown in Figure 7.10. You can add basic colors to be detected as well as choosing specific colors from the current image. After you have selected colors (or a variety of shades of the same color) you wish to track, adjust the parameters in the lower-right corner of the window so that only the object you wish to track can be seen in the video window of Figure 7.9.

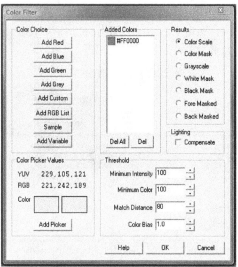

Figure 7.10: Dialog box for RoboRealm's color filter.

After your adjustments allow the desired object to be seen as you move it within the camera's view you are ready to add another module. Expand the **Blobs** option in the left column and select **Blob_Size** to expose the window shown in Figure 7.11. You can make adjustments so that only groups of pixels (blobs) of a specified size and intensity are displayed. At this point, the video display should show only the movement of the object you wish to track. If necessary, you can fine-tune your options in each of the modules.

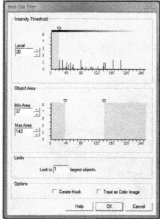

Figure 7.11: This dialog allows you to limit the
blob size to be recognized.

Next, expand **Analysis** and select **Center_of_Gravity**.
You should see the window shown in Figure 7.12. This
module will essentially find the center of a blob of pixels
and insert statistics about that blob (such as its **x,y**
coordinates) into a number of predefined variables. A box
will be drawn around the blob and the coordinates
displayed. You can choose to have this information
overlaid on the source images as shown in Figure 7.13.

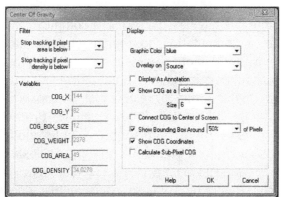

Figure 7.12: This dialog instructs RoboRealm to provide
information about the object being tracked.

If everything is adjusted properly the image will be tagged with coordinates representing its position on the screen when it comes in view. As you move the object the box identifying its position should travel with the object and the coordinates should constantly reflect the current position. Figure 7.13 shows RoboRealm tracking the red end of a pen.

Figure 7.13: RoboRealm provides the coordinates of the object being tracked.

At this point, we have used a series of complex mathematical computations to implement a fairly robust vision algorithm without having to understand or program any of the mathematics. Even though this is a very usable algorithm, it is important to realize that the capabilities of RoboRealm far exceed this simple example. Properly utilizing some of RoboRealm's more complex modules, for example, should allow you to identify and track shapes instead of colors and even perform facial recognition.

The RoboRealm library is so extensive that a new user can easily be intimidated. However, with experimentation and exploration of the RoboRealm examples, help files, and forum posts, you should be able to implement vision algorithms that exceed anything you could have done on your own. RoboRealm also has the ability to directly control motors, so a complete robot can be created using RoboRealm alone. Many applications need the features of a specific language though, so it is useful for the information generated by RoboRealm to be transferred to other programming languages.

Fortunately, RobotRealm has provided many ways to allow the information it generates to be shared with other programs. For example, the data could be sent over a serial link to a microcontroller. While this is an excellent option, it is doubtful that a microcontroller could fully exploit RoboRealm's capabilities. In order for a robot to fully utilize vision it must be controlled by a full-featured high-level language capable of complex mathematical operations, large multi-dimensional arrays, and many other features not found in the microcontrollers typically used to control hobby robots.

Let's assume RobotBASIC is running on the same machine as RoboRealm. Our goal is to demonstrate generic ways of interfacing RoboRealm with RobotBASIC so that the vision information can be obtained and utilized to make high level decisions.

The first interfacing method we are going to examine is the least sophisticated, but because of its simplicity, it should work, not just with RobotBASIC, but with any high-level language. RoboRealm provides a **VBscript** module that we used to format the **x,y** coordinates of the blob being tracked into a single ASCII string. To make it easy for the receiving program to extract the data, we formatted the information so that it is sandwiched between the letters **x**, **y**, and **z**. For example, if the blob is located at coordinates 57,109 then the data will be formatted like this: x57y109z.

Once our desired information was formatted, we used RoboRealm's **Keyboard_Send** module to insert the data into the PC's keyboard buffer. This module allows you to specify which application should receive the keystrokes.

The format of the data makes it easy for the receiving program to watch the stream of keystrokes coming from RoboRealm and convert the coordinates to integers. Figure 7.14 shows how easily this can be accomplished with RobotBASIC. The details of the conversion are encapsulated in a subroutine called `GetCoordinates`. After the routine is called, the two variables in the parameter list will contain the **x,y** coordinates of the blob being tracked.

The subroutine is more complex than necessary because it uses a timer to set the coordinates to $(-1,-1)$ and return control to the main program if data has not been received within 1000ms. If the object goes out of view, RoboRealm will send zeros for the x,y coordinates. This means that any values greater than zero can be considered valid.

The logic of this example should be easily implemented in other languages, making it effortless for anyone to experiment with vision. You do not have to understand any of the complexities of the RoboRealm program. Simply download the program and edit the **Keyboard_Send** module so that it sends keystrokes to your program. Note: RoboRealm versions 2.30.6 and later provide greater control over which program receives the keystrokes. You can, for example, select RobotBASIC's terminal output window instead of just RobotBASIC. With earlier versions you might have to tell RoboRealm to force focus on the intended application.

```
Main:
  while true
    call GetCoordinates(a,b)
    xyString 100,100,a;b,"   "
  wend
end

sub GetCoordinates(&x,&y)
  t=timer()
  x="" \ y=""
  repeat
    getKey k
    if timer()>t+1000 then x=-1\y=-1\return
  until char(k)="x"
  while true
    waitKey k
    if char(k)="y" then break
    if k<>0 then x += char(k)
  wend
  while true
    waitKey k
    if char(k)="z" then break
    if k<>0 then y += char(k)
  wend
  x=ToNumber(x,-1)
  y=ToNumber(y,-1)
return
```

Figure 7.14: This program extracts the screen coordinates from a keyboard stream.

You can use this information to control the simulated eyes just like we did in the previous program. Full source code for this additional example program can be downloaded from our web page.

Alternative Communication

RoboRealm also has a module that can send data to other applications over the Internet or a LAN. Since RobotBASIC has commands for Internet communications, we used that module to create our next example. An RB program that can capture the UDP data is shown in Figure 7.15. The program starts by printing your computer's IP address so you can use it to setup the RoboRealm **Socket_Communications** module. The translation portion of the program in Figure 7.15 differs from the keystroke program in Figure 7.14 because the UDP data is captured as a string instead of character by character.

```
Main:
  print "Local IP address = ",TCP_LocalIP()
  UDP_Start("MySocket",4040)
  while true
    call GetCoordinates(a,b)
    // print if object is in view
    if a<>-1 then xyString 100,100,a;b,"  "
  wend
end

sub GetCoordinates(&xCoord,&yCoord)
  xCoord = -1 \ yCoord = -1
  if UDP_BuffCount("MySocket")>0
    d=UDP_Read("MySocket")
    px=InString(d,"x") \  py=InString(d,"y")
    xCoord = ToNumber(SubString(d,px+1,py-px-1),-1)
    yCoord = ToNumber(SubString(d,py+1),-1)
  endif
return
```

Figure 7.15: RobotBASIC commands make it easy to obtain and extract the screen coordinates from a UDP message.

Summary

In summary, it is easy to implement powerful vision algorithms with RobotBASIC alone and if you need even more power consider utilizing RoboRealm in your projects.

Chapter 8

Communications and Control Over the Internet

A major feature of RobotBASIC is its ability to control remote robots over a wireless link such as Bluetooth or Xbee. If you are not familiar with this capability, refer to the rCommPort command in the HELP file. You might also consider some of our advanced interface books available on our web page. The material in this text provides much of the background needed to be able to better utilize our advanced books.

Typically, when a robot is controlled over a wireless link, the robot is relatively close to the computer running RobotBASIC. This chapter details how to control a robot, or other devices, from anywhere in the world. The example programs provided in this chapter were tested between the United States and Australia - you can't get two computers further away from each other and remain on the planet.

Internet Communication

In order to get this kind of control, the two computers in question will communicate with each other over the Internet. Writing programs to handle this type of communication is generally beyond the skills of average hobbyists, which is why we created RobotBASIC commands and functions that can handle the complex details for you.

RB's internet-related commands are so simple that most readers should be able to understand and use them by simply studying the two programs that will be provided in this chapter. The whole point is that you can use these commands without having to understand most of the details of how they work, or even how the Internet itself works.

That said, we know that many readers will want more information about these commands and how they can be used even more effectively than we will show in the simple example applications. Some readers will even want more information about the Internet itself. For such readers, we have provided two extensive Appendices (C and D) that provide exactly these kinds of details. For everyone else, let's just see how easy it is to create an interesting application that requires bi-directional communication over the Internet.

Controlling a Simulated Robot Over the Internet

The application developed in this chapter will consist of two programs that are connected only by a LAN or the Internet. Essentially, the user will be able to use the arrow keys on one computer to control a robot on another.

The user program (the client) will gather keyboard input from the user and transmit that information to the second program (the server) that will use it to control a simulated robot. The second program will gather sensory data from the simulation and send it back to the first program.

When the first program receives the sensory data, it will display it (see Figure 8.1) so that the user can easily visualize what sensors are being triggered. The circles around the robot represent the IR proximity sensors and the boxes represent the bumpers. When a sensor is triggered, the boxes and circles are filled in.

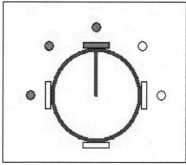

Figure 8.1: This display lets the user see the status of bumper and IR proximity sensors.

Figure 8.2 shows the server program and 8.3 shows the main portion of the client program. Notice they both start by printing the computer's IP address. Use the addresses printed to change the addresses in your programs to reflect the address of the other machine (read the comment in each program).

```
//Server:receives keys, moves robot, and
//returns sensor data
print TCP_LocalIP()
Client_IP = "192.168.1.100"
//change above to what is on the Client screen
Client_Port = 55000
UDP_Start("Server",54000)
//rCommPort 17
rlocate 100,100
while true
   if !UDP_BuffCount("Server") then continue
   x = getstrbyte(UDP_Read("Server"),1)
   if x=kc_UArrow  then if !(rBumper()&4) then rforward 1
   if x=kc_DArrow  then if !(rBumper()&1) then rforward -1
   if x=kc_LArrow  then rturn -1
   if x=kc_RArrow  then rturn 1
   if x=32 then rForward 0
   s = char(rFeel())+char(rBumper())
   UDP_Send("Server",s,Client_IP,Client_Port)
wend
end
```

Figure 8.2: The server program receives user data to control the robot and then returns sensor data.

The programs use the port addresses 54000 and 55000 because they are generally unused on most machines. See Appendix C for more information on this topic.

When the server's buffer contains data, it is obtained using the `UDP_Read` command and compared to predefined constants for the arrow keys. Notice that the robot moves based on the keystrokes received, but only if the bumper sensors do not indicate that an object has been detected.

```
//Client Program: sends the keys, receives IR and
//Bumper data
print TCP_LocalIP()
gosub InitSensorDisplay
xyString 50,150,"Use Arrow Keys to move robot (space
bar to update sensors without movement)"
Server_IP = "192.168.1.100"
//change avove as is displayed on the server screen
Server_Port = 54000
UDP_Start("Client",55000)
allowed=char(kc_UArrow)+char(kc_RArrow)+\
   char(kc_LArrow)+char(kc_DArrow)+" "
while true
  K = char(KeyDown())
  if InString(allowed,K)
     UDP_Send("Client",K,Server_IP,Server_Port)
     repeat
     until UDP_BuffCount("Client") >=2
     x = UDP_Read("Client")
     IR = GetStrByte(x,1)
     Bump = GetStrByte(x,2)
     xyString 0,20,"Ir    = ",IR,"    "
     xyString 0,40,"Bumper = ",Bump,"    "
     call ShowSensors (IR,Bump)//(IR,BUMP)
  endif
wend
```

Figure 8.3: The client program sends keystroke data and receives sensor information.

The client program in Figure 8.3 ignores all keystrokes entered by the user except for the arrow keys (which are sent to the server program). When sensor data is received, it is displayed as shown in Figure 8.1. Figure 8.4 shows the subroutines needed to create this display. The graphic statements used are easy to follow. Consult the HELP file if you are unfamiliar with any statements.

```
InitSensorDisplay:
  x=400   \ y = 300 // position of display
  r=50              // radius of displayed robot
  SetColor Blue \ LineWidth 4
  Circle x-r,y-r,x+r,y+r
  Line x,y,x,y-r
  // Setup Bumper positions
  Data BumpX; x,x+r+5,x,x-r-5
  Data BumpY; y+r+5,y,y-r-5,y
  Data Mask; 1,2,4,8,16
  LineWidth 1
  for i=0 to 3
    if not(i#2)
      rectangle BumpX[i]-15,BumpY[i]-3,\
             BumpX[i]+15,BumpY[i]+3,black
    else
      rectangle BumpX[i]-3,BumpY[i]-15,BumpX[i]+\
             3,BumpY[i]+15,black
    endif
  next
  // Setup IR positions
  Dim IRX[5],IRY[5]
  for i=0 to 4
    IRX[i]=x+(r+25)*Cos(dtor(45*i))
    IRY[i]=y-(r+25)*Sin(dtor(45*i))
    Circle IRX[i]-5,IRY[i]-5,IRX[i]+5,IRY[i]+5,black
  next
return
//-------------------------------------------
Sub ShowSensors(IR,Bump)
  LineWidth 1
  for i=0 to 3
    if Bump&Mask[i]
      Scolor=Red
    else
      Scolor=White
    endif
    if not(i#2)
      rectangle BumpX[i]-15,BumpY[i]-3,BumpX[i]+\
             15,BumpY[i]+3,black,Scolor
    else
      rectangle BumpX[i]-3,BumpY[i]-15,BumpX[i]+3,\
             BumpY[i]+15,black,Scolor
    endif
  next
```

```
for i=0 to 4
  if IR&Mask[i]
    Scolor=Red
  else
    Scolor=White
  endif
  Circle IRX[i]-5,IRY[i]-5,IRX[i]+5,\
          IRY[i]+5,black,Scolor
next
return
```

Figure 8.4: Add these routines to the client program in Figure 8.3 so it can display the sensor data.

LAN or Internet

The two programs in this chapter can be easily run on two computers (or even two copies of RobotBASIC running on the same machine) over a LAN (e.g. your home wireless network). They can also be run on two machines located on totally different networks connected only by the Internet. When communicating over the Internet though, you must prevent firewalls from blocking the data transfers. Refer to the Appendix C for more information on this topic.

Summary

The programs in this chapter make it easy to see how data can be communicated over the Internet. For many people that may be all that is needed. For more detailed information refer to Appendices C and D.

Chapter 9

What's Next

This book provides the fundamental principles of controlling external devices with RobotBASIC but this is only the beginning. Everything you have learned here is applicable to standard control programs, and to the way robots are typically controlled.

Why RobotBASIC?

Since programs written in RB run on a PC (instead of on a microcontroller) where there is plenty of processing power and memory, users get the benefit of unlimited multidimensional arrays, floating point variables, and a plethora of unique commands and functions that handle many tasks that are difficult, if not impossible, to achieve in microcontroller languages.

When it comes to developing algorithms for robotic behaviors, perhaps one of RobotBASIC's most powerful tools is an integrated robot simulator. While the two-dimensional simulated robot may not seem realistic at first, the variety of sensors on the robot and the methods with which they are used and programmed provide a realistic life-like experience. The robot has bumper sensors, perimeter proximity sensors, line sensors, ranging sensors, a beacon detector, an electronic compass, a battery-level sensor, and a simple GPS system.

Debugging code on a real robot has always been challenging. Faulty programs can easily damage the robot itself and the erratic operation of untested sensors can make debugging a new algorithm a nightmare. RobotBASIC's simulated robot and integrated debugging tools address both of these problems.

A New Paradigm in Robotics

You may presume that RobotBASIC's simulation capability is only valuable as a prototyping tool. You can certainly develop and debug a behavioral algorithm using the simulation, and then translate the principles associated with the finished program into the native language of your preferred microcontroller, but RB provides a new approach for controlling your robot - *a new paradigm for hobby robotics*.

It is worth noting that many industries have successfully used the principle of *simulate then deploy* for decades so it makes sense that this approach be applied to robotics. Microsoft's Robotic Studio (RS) also follows this paradigm, but its complexities tend to make it more suitable for professional developers than hobbyists. One reason RS is so complex is that it supports an endless variety of configurations. RB achieves simplicity by limiting the robot's sensory system to a proven set of sensory types and placements. Nevertheless, if more power is needed there are commands that can allow you to also create your own arrangement of additional sensors. Our book *Robot Programmer's Bonanza* demonstrates that the sensors arrangements chosen for RB are suitable for achieving many robotic behaviors.

RobotBASIC has a built-in protocol for controlling a *real* robot using the *very same* programs used to control the simulated robot. The control can extend over a Bluetooth or XBee (or other) wireless link for small robots, or a Netbook computer running RobotBASIC can be embedded in larger robots. Either method lets you program your robot using a full-featured language with virtually unlimited variable space for integers, floats and strings. You will have access to multidimensional arrays, trigonometric functions, file I/O, and a host of other tools that can make programming a robot easier than ever before.

Our book, *Enhancing the Pololu 3pi with RobotBASIC* shows both the hardware and software details of how most of RobotBASIC's simulated sensors can be implemented on a real robot. Figure 9.1 shows a photo of the standard 3pi robot from Pololu and Figure 9.2 shows a modified 3pi fully loaded with sensors that correspond to the simulated sensors available on RB's simulation.

Figure 9.1: This unmodified 3pi robot served as the basis for building the robot in Figure 9.2

Figure 9.2: This robot has nearly all the sensors as RobotBASIC's simulated robot.

Let's examine the basic principles used by RB to control the modified robot. When an RB program wants to move the simulated robot, it might issue a command such as rForward 20. (All robot commands in RB start with the letter r.) This command normally moves the simulated robot forward 20 pixels, which is default radius of the robot. A command such as rTurn 10 would turn the robot 10° to the right. Functions such as rFeel() and rBumper() interrogate the simulated sensors on the robot and return values that correspond to the current environmental situation.

These types of commands and functions make it easy to control the simulated robot with appropriate algorithms.

To have these very same commands and algorithms control a real robot, you simply add the command

```
rCommPort N
```

at the beginning of the program. This command indicates that a Bluetooth adapter (or other serial communication device) is attached and using serial port **N** (typically a virtual port).

RobotBASIC's Internal Protocol

Once the rCommPort command has been issued, all RB's robot-related commands and functions (including rForward and rTurn) no longer control the simulation. Instead, they *automatically* send two bytes to the specified serial port. The first of these bytes is an op-code that identifies the command. The table in Figure 9.3 shows the op-code used for each of the simulator commands implemented on the ***enhanced*** 3pi robot.

Command	Op-code
RLocate	3
RForward	6
(backward)	7
RTurn (right)	12
(left)	13
rCompass	24
rBeacon	96
RRange (right)	192
(left)	193
RSpeed	36
rChargeLevel	108

Figure 9.3: RobotBASIC's internal protocol uses these op-codes to communicate with remote robots.

The second byte sent to the robot is either zero, if not needed, or a parameter related to the command. For example, when the command rForward 20 is issued, the PC will send out a 6 followed by a 20. The 6, of course indicates the FORWARD operation is being requested and the 20 specifies how far forward to move. The units for the real robot obviously cannot be pixels, but we must maintain compatibility if the behavior of the real robot is to mimic the behavior of the simulation. Since a movement of 20 pixels represents the radius of the simulated robot, the code running on the real robot should move it a distance equivalent to the real robot's radius when an rForward 20 command is received. All movement commands implemented on the real robot should respond proportionally in this manner.

In addition to receiving commands from the PC, the programming implemented on the microcontroller of the external robot has to return sensory data to RB. RobotBASIC's communication protocol requires that the robot return five bytes of sensory data *every* time it receives a command, thus ensuring synchronization. Three pieces of data (information from the bumper sensors, the perimeter object detection sensors, and the line sensors) are very time sensitive and are nearly always returned in the first three of these five bytes (in the order listed). Data from these sensors is automatically extracted by RB and used appropriately whenever a program asks for it by using the functions rBumper(), rFeel(), and rSense().

The remaining two bytes of the returned data are usually zero because they are typically not needed. When functions such as rCompass(), rBeacon() and rRange() are executed though, these two bytes are used to return the requested data.

Remember, the above functions normally return the state of the simulated robot's sensors. After an rCommPort command has been issued though, each of these functions will return the status of the real robot's sensors. Let's look at the example program in Figure 9.4 to help clarify these ideas. Notice the first line of the program is commented out for now.

```
//rCommport 40
rLocate 400,300
while true
  while rFeel() = 0
     rForward 1
  wend
  rTurn 150+random(60)
wend
```

Figure 9.4: This sample program moves the robot randomly through its environment.

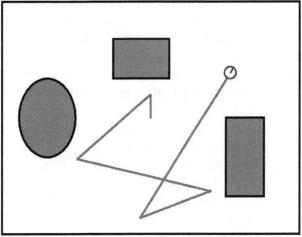

Figure 9.5: The program in Figure 9.4 will move the robot similar to this screenshot.

The program starts by initializing the simulated robot at position 400,300 on the output screen. Next, an endless `while`-loop repeats the heart of the program where an inner loop moves the robot forward as long as it does not *feel* an object in its path. When an object is detected, the robot turns away from it a random amount and the program continues. Figure 9.5 shows the path the simulated robot might take through a sample environment.

When the first line is un-commented allowing the `rCommPort` command to be executed, the program will automatically control the 3pi robot causing it to perform like the simulated robot - that is it moves until it encounters an object, then turns away. As long as the robot's behavior is determined by sensory data, as in this example, the real robot will respond like the simulation. Examples of such sensor-controlled behaviors are following a line, hugging a wall, finding a doorway, tracking a beacon, and negotiating a maze.

Advantages of the New Paradigm

There are many advantages when programming a robot in accordance with the new paradigm. First, the user gets to program in a powerful full-featured environment with an English-like syntax instead of a cryptic microcontroller language. Of course, the remote robot must be initially programmed to handle the communication exchanges between itself and RB, but this has to be done only once. Furthermore, the majority of the programming for the microcontroller is limited to reading sensors and motor control. The complex algorithms involving the *intelligence* of the robot can be implemented in the high-level language, making the development cycle easier by far.

There are no files to compile and nothing to download. If several people in a robot club or school classroom have RB compatible robots, they can easily share entire programs, or even libraries of routines of behaviors that they develop with their robot or just using the simulator alone. ***And this sharing is possible even if each of the robots uses totally different microcontrollers.***

The embedded code for different robots might be totally different of course, but each robot will respond in the same manner to the algorithm implemented by the RobotBASIC program. If you are interested in how the external robot is programmed, the source code for the 3pi robot is available for free download at www.RobotBASIC.com. It is relatively complex though, so the accompanying book is recommended reading. Our more advanced book, *A Hardware Interfacing & Control Protocol*, explores the principles of the new paradigm more in depth, developing a methodology for applying it more rigorously on a Parallax Propeller microcontroller.

It is important to realize that the robots being controlled through this approach can have a variety of sizes and use different kinds of sensors, motors and microcontrollers. As long as appropriate sensors are chosen and mounted like those on the simulated robot, the embedded firmware can isolate application programmers from nuances that do not concern them, thus enabling them to spend more time on algorithmic development.

Remember the example in Chapter 8 where a simulated robot could be controlled over the Internet. By adding one rCommPort command to that system, the real robot can be controlled instead of the simulator (we did it using the modified 3pi). This is a powerful concept and is further expounded in the advanced book with additional practical and interesting projects.

This approach to building and controlling a robot may not fit everyone's needs, but it provides advantages that are worth examining. The RobotBASIC team is dedicated to making it easy for hobbyists to build their own RB compatible robots. Principles learned from the 3pi project are already being used to design a RobotBASIC Robot Operating System (RROS) that is expected to run on a Parallax Propeller-based robot controller board.

Once the controller board and the RROS are finalized, motors and sensors of various sizes and types from a variety of manufacturers can be connected and utilized with only a minimal knowledge of electronics. The RROS will automatically control the motors and gather sensor data, mapping it to the RobotBASIC functions so that the entire process is seamless, making programming a real-life robot no harder than programming the simulated one.

Summary

Now that you understand the principles of hardware interfacing, its time to test your skills by developing your own applications. When you are ready, try building a robot based on our new Paradigm. Our books *Enhancing the Pololu 3pi with RobotBASIC* and *A Hardware Interfacing & Control Protocol* carry on where this one leaves off with more projects and hardware examples that show strategic methodologies for implementing the new paradigm at more advanced levels.

Appendix A

Bit-wise Operations

Most programmers are familiar with *logical* AND and OR operations but if you are not hardware oriented you might benefit from a short tutorial on *bit-wise* operations. The goal of this appendix is to assist you in the programming examples used elsewhere in the text. It should not be considered a complete course on binary mathematics.

Logical Operations
RobotBASIC uses AND and OR to represent logical operations. You can also use the C-style operators && and ||. Logical operations are used for testing true/false conditions. For example, you might test if a number is greater than 100 and less than 200 like this.

```
IF x>100 AND x<200
```

Bit-wise Operations (2 Inputs)
Bit-wise operations are used to examine and alter the individual bits (the ones and zeros) in a number. This is essential for hardware interfacing. In order to understand bit-wise operations, you need to fully understand logical operations because bit-wise operations are just the logical operation applied to each bit position.

The Table in Figure A-1 shows the outcomes for AND, OR, and XOR operations. The AND condition is only true (1) if both inputs are true while the OR condition is true if either input is true. The exclusive-or operation (XOR) is true if either input is true, but not both.

Inputs		AND	OR	XOR
0	0	0	0	0
0	1	0	1	1
1	0	0	1	1
1	1	1	1	0

Figure A-1: These output conditions apply to both logical and bit-wise operations.

Hardware gates perform bit-wise operations for you. For example, the only way the output of an **AND** *gate* will be high (typically 5 volts) is if both of its inputs are high. It is often valuable to perform such operations in software rather than use hardware gates. Let's look at an example.

Assume that we have three switches or sensors connected to an input port and we have read the three bits of information on the port (**A**, **B**, **C**) of that port into the variable X. We might want to perform different operations depending on which of the three switches (or sensors) is active. Figure A-2 shows the binary numbers associated with each of the possible input combinations available from a 3-bit port.

Decimal	Binary		
X	**A**	**B**	**C**
0	0	0	0
1	0	0	1
2	0	1	0
3	0	1	1
4	1	0	0
5	1	0	1
6	1	1	0
7	1	1	1

Figure A-2: Decimal and binary numbers are just different ways of representing the same information.

For example, let's assume we wanted to do something if the most significant bit position (**A**) is high. If we did it like this:

```
IF X=4
```

the statement would be true, but only if the other bits were zero. If we want the condition to be true if the MSB is a one regardless of the condition of the other two bits we might use a statement like this:

```
IF X>=4 AND X<=7
```

While this statement performs as desired, it is cumbersome and becomes even more so as the complexity of the decision increases. A better way to handle this situation is to use bit-wise operations. As with logical operations, RobotBASIC allows either BASIC or C-style syntax for bit-wise operations as shown below.

Operation	BASIC Syntax	C-style Syntax
Bit-wise AND	BAND	&
Bit-wise OR	BOR	|
Bit-wise XOR	BXOR	@

Figure A-3: RobotBASIC offers two syntax options for many operations.

We can use the bit-wise AND operation to examine the value of a specific bit (or bits) in a number. If, for example, we wanted to look at only the MSB, we could BAND the data being examined with 4. Let's see how this works.

Assume that the data to be examined is contained in the variable X and that its binary equivalent (just 3 bits for simplicity) can be represented by **A**, **B**, and **C**. Consider this statement:

IF X&4

What this really means is that *each bit* of the variable X is individually ANDed with the corresponding bit in the number 4. Since 4 is 100 in binary, this bit-wise ANDing can be represented as shown below.

Bits of X	**A**	**B**	**C**
Bits of 4	1	0	0
Result of bit-wise AND	**A**	0	0

Notice that when **A** is ANDed with 1, it assumes the value of **A**. Think about it. If **A** is 0 then the result will be 0, and if **A** is 1 then the result will be 1 (use Figure A-1 to confirm this). The output from the lower two bit positions will always be 0, because anything ANDed with 0 will be 0.

Using similar logic, we can test to see if either bit **A** or bit **C** is a 1 with the following statement.

IF X & 5

Remember, the IF statement will be true as long as the expression results in a non-zero value. In this example, X&5 will be non-zero if either or both of the bit positions **A** and **C** are 1's. Other bits in the number will have no effect on the decision. RobotBASIC allows you to specify the *masking* number in binary which can make the desired operation more obvious. This statement, for example, provides the same operation as the one above, because 0x101 is the binary equivalent of 5.

```
IF X & 0x101
```

Clearing Bits

We can also use bit-wise AND operations to *clear* bits in a number. To do so, simply BAND the bits you want cleared with 0. For example, the following statement will clear the two LSBits in a number stored in the variable X.

```
X = X & 0x11111100
```

Setting Bits

The bit-wise OR operation can be used to *set* bits in a number, because any bit ORed with a 1 will result in a 1. The following statement will set the two least significant bits in the variable X. You could of course use the binary number 0x11 in this example (or use BOR instead of |).

```
X = X | 3
```

Toggling Bits

The bit-wise XOR operation (refer to Figure A-1) can be used to toggle a bit from its current state. This means that a 0 changes to a 1 and a 1 changes to a 0. The following example reverses the values of the most-significant bit and the least-significant bits in an 8-bit data stored in the variable X.

```
X = X @ 0x10000001
```

Unary Operations

Operations can be performed on a single number too. For example, you can apply a logical NOT as shown below. Both lines do exactly the same thing.

```
a = NOT b
a = !b
```

In this case a will be 0 as long as b is anything but 0 and a 1 if b is zero. A bit-wise NOT is done as shown below.

```
a = bNOT b
a = ~b
```

In this case a will assume the value b with every bit reversed. For example, if b is the binary number 00001101 then a will be 11110010.

Refer to the RobotBASIC HELP file for more information on these and other bit-wise operations.

Appendix B

Finding Serial Port Numbers

When you add devices such as USB serial ports and USB Bluetooth adapters, Windows sets up a virtual serial port that makes the device appear as if it is a standard serial port. This makes it easy for RobotBASIC applications to interface with these devices using the standard serial commands. To use these commands though you need to initialize the communication using `SetCommPort` (or `rCommPort` for the built-in simulator protocol). This means you need to know the port number assigned to the virtual port by Windows. There are several ways to obtain the port number depending on the device you are using, but this Appendix shows a simple method that should work with all devices.

This Appendix assumes you have installed the device in question. In most cases, just plugging in the device will typically activate a plug-and-play session that will install standard Window's drivers for you. In the case of Bluetooth adapters, you will need to pair the adapter with the remote transceiver. Refer to the documentation for your specific hardware.

In order to find the port number for your device, make sure your device is plugged into a USB port then open the Window's Control Panel and select the System icon. From there select the Device Manager.

In the Device Manager window, find the **Ports** entry and expand it by clicking the + sign. You should see the serial and parallel port devices that are active on your system. Find the serial device you are using and make note of the port number assigned to it. Use that number in the SetCommPort command as indicated in the examples throughout this text.

Using RobotBASIC to Find Out the Available Serial Ports

The command SerPorts in RB returns a string with all the available serial port numbers separated by Cr/LF. The following program will list all the ports that are active serial ports:

```
SerPorts P
if P != ""
  print "Available ports are:",crlf(),P
else
    print "There are no valid serial ports"
endif
```

The program does not tell you what device is on what port, however.

Appendix C

Utilizing the TCP and UDP Protocols

obotBASIC has a very easy to use interface that facilitates a communication link between networked computers over a Local Area Network or through the Internet. With just a handful of functions you can implement a bidirectional communication conduit between two or more networked computers using either the Transmission Control Protocol (TCP) or the User Datagram Protocol (UDP). We have found most hobbyists are not familiar with the details of LAN/Internet communication so this appendix tries to provide some information.

With RobotBASIC's collection of TCP and UDP functions even the novice programmer can easily implement a whole range of interesting projects that would otherwise challenge an expert even with advanced development tools.

Imagine being able to collect instrumentation data on a PC in Australia and sending the data to a machine in the USA for display purposes and for parameter settings and so forth.

Imagine being able to text chat with a friend in Australia while you are in the USA. Yes, of course you can do that on numerous web sites, but think of the pleasure of writing the program that can do this, yourself.

Scope
This appendix deals with the details of how to use RobotBASIC's functionalities to send data between networked computers. In order to make this appendix more effective, a number of terms will be defined and some background information will be provided.

Some Terminology
Throughout the appendix there will be reference to certain terminologies that pertain to the field of networking. It is necessary to have a working knowledge of what these terms signify to be able to effectively utilize the functionalities provided in RobotBASIC. The given explanations are for practical use and are not meant to be an in depth explanation.

There are numerous ways of implementing a network using a plethora of hardware. It is not feasible to cover all possible combinations, so only the setup shown in Figure C-1 will be considered. If your network differs from the arrangement shown you will need to consult with a network administrator if you wish to communicate across the Internet so as to handle firewalls and other issues. However, if you will be confining your projects to communicate computers inside the same LAN then the information in this appendix is all you need. Later on we will discuss how to configure the arrangement in Figure C-1 to enable communication between two PCs across the Internet.

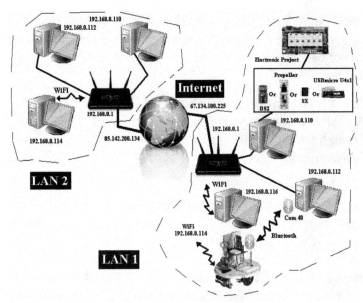

Figure C-1: A Typical Network Arrangement.

In Figure C-1 notice that one of the computers is connected to electronics. This is to indicate that any or all of the computers on the network can be performing data acquisition or control through external instrumentation. Furthermore, any or all of the computers can carry out additional channeling of communicated data over a wireless radio or through Bluetooth to a robot or any other microcontroller performing instrumentation. Additionally, you can have any number of autonomous robots that have their own WiFi TCP or UDP abilities and thus can be communicated with directly from any computer on the network (not through a secondary channel) just as if they were a computer on the network, which in fact, they would be.

A Local Area Network (LAN)
A LAN is a group of computers dispersed over a relatively small area such as a home or office building. The machines are interconnected using physical wires or wirelessly using

some interconnection system. Usually, there is a central server that serves as the communications coordinator. This server can be a computer or a specialized device called a *Router*.

An Internet Service Provider (ISP)

An ISP is a company that provides a computer system through which clients can obtain connectivity between their LANs and the Internet.

The Internet

The Internet is a very large and complex network of interconnected ISPs around the world. An ISP provides a method for a computer in one LAN to be able to communicate with another computer in another LAN by managing the routing through other ISPs until eventually reaching the ISP of the remote LAN and then on to the final target computer.

A Router

A Router is a device that manages the connectivity of the various machines in the LAN. Computers in the LAN can be connected to the router through a physical wire or by *WiFi*. A router is sometimes referred to as a NAT (Network Address Translator) and often serves as the connection point between the LAN and the ISP which then provides a link to the Internet. Additionally, the Router often acts as a *Fire Wall* as well as a DHCP (Dynamic Host Control Protocol).

WiFi

WiFi is a system of hardware and software for linking a computer to the LAN wirelessly. For all intents and purposes the computer will appear as if it is connected to the rest of the LAN over a wire. Other than the convenience of mobility and lack of cumbersome wires the computer is no different, from the network's point of view, than a wired computer.

Fire Wall

A Fire Wall is software and/or hardware used for blocking access from outside a LAN to the computers inside the LAN. This is a safety measure to prohibit illegitimate access to the computers in a LAN. A Fire Wall can also limit access from inside the LAN to the outside. You can configure a Fire Wall to allow certain communications while blocking others.

Dynamic Host Control Protocol (DHCP)

A DHCP, in short, assigns an IP address to each computer on the LAN.

Network Address Translator (NAT)

A NAT makes sure that computers inside the LAN appear to the outside world as valid computers with valid global IP addresses even though their actual IP addresses are only local addresses with significance only inside the LAN. An IP Address is basically the name of the computer. It is a set of 4 numbers separated by a dot (e.g. 192.168.0.120), however the whole address is not a number; rather it is a text. Each individual number in the 4 fields ranges from 0 to 255. The network will not function without some method of addressing a particular computer and the IP address is this method.

Socket

A socket is the endpoint of a bidirectional communication flow across the network. When data flows between two computers or when a connection is established between them it is achieved through a complex program called a Socket. This program is part of the internal structure of RobotBASIC with which you interact through a handful of function calls. To carry out TCP communications you do it through a TCP Client Socket (TCPC) to and from a TCP Server Socket (TCPS);see later for more details. Communications over the UDP are between two UDP Sockets (see later for more details). A Socket is identified by the IP address of the machine it is running on and a Port number. The port number must not be previously assigned to any other socket (UDP or TCP) on the same IP address. A Port Number is a further subdivision of a particular IP address. A machine has one IP address but it may have various Sockets (programs) that provide access to the LAN. These Sockets will be addressed on each machine through an additional number that can be considered as a sub-address on the machine. This number is called the Port number and is an actual 16-bit number that ranges from 1 to 65535 (0xFFFF). Think of an IP address as the street address of a building and the port number as the number of particular apartment in the building. The building is the computer and apartments are the various Sockets (programs) running on the computer. This is why a port number associated with a socket must not be in use by any other active socket.

Transmission Control Protocol (TCP)

TCP is a sophisticated standard for moving data over a network. There is no need to understand this standard in depth in order to use the functions in RB. TCP is a client-server protocol. A server socket can accept connections from multiple client sockets but a client socket can only be connected to one server socket at a time. Once a client connects to a server it becomes as if there is a direct wire between them and data can be exchanged between the two sockets. See later for more details.

User Datagram Protocol (UDP)

UDP is a simpler standard than TCP for exchanging data over a network. It is not necessary to know the details of the protocol to be able to use the functions in RB. UDP is not a connection oriented protocol. A UDP socket can send to any other UDP socket without establishing a link first. There is no exclusive or maintained connection between the two machines. See later for more details.

Simple Mail Transfer Protocol (SMTP)

SMTP is a protocol that ensures that an email is able to reach the intended recipient's computer from the sender's computer through the ISP of the sender around the Internet and finally to the ISP of the recipient and on to the recipient's computer.

Remote IP Address

A remote IP address is the address of the machine that is hosting the destination socket. That is the remote socket to which the data is going to.

Local IP Address

A Local IP Address is the address of the machine that is hosting the originating socket. That is the socket being used to send data.

Packet

A Packet is a set of bytes (buffer or string) that is sent in one send action using TCP or UDP. When you send data over the protocol you are sending multiple bytes in one go. This is called a packet. A big amount of data can be sent in multiple chunks (Packets). UDP has a limit on the size of a packet (2048 bytes) while TCP does not.

Utilizing the UDP System:

In RobotBASIC you can create multiple UDP sockets in a program. Each socket must be assigned a unique and not currently in use Port number. When created, the socket will *automatically* use the IP address of the machine it is created on, but, you must assign it a particular port number.

> ⚠You *must* ensure that the port assigned to a UDP socket is unique and is not being used by another active socket (UDP or TCP) in the same program or other programs on the same machine.

To create and activate a UDP socket use the command
`UDP_Start(se_Name{,ne_ListenPort})`. Notice that the
port number is optional. If you do not specify a port
number it will be assumed to be 50001.

Another parameter of the function is the socket name.
When you start a UDP socket you assign it a name. This
name is then used to refer to the socket in further functions
that utilize it. *The name therefore has to be unique (within
a program not across other programs) and is case
sensitive.*

Once a socket is started you can begin to use it to
communicate with other UDP sockets. These other sockets
can be on the same machine (IP) or on a remote machine.

> ⚠You must not try to communicate with a non-existing socket on
> the same machine (IP) as the socket you are using. It is not a
> problem if you try to communicate with a non-existing socket on a
> remote machine, data will just not get there since the socket does
> not exist. However, there is a problem with the Windows 2000 and
> Windows XP operating systems (not others) where if you try to
> send data to a socket that does not exist on the same machine as the
> sender socket the OS hangs. Therefore, *Do Not Send Data To A
> Non-Existing Socket.*

The functions discussed in the following sections enable the
writing of a program to send and/or receive data packets over the
UDP. A socket in a RobotBASIC program can communicate
over the LAN or Internet with another socket that does not have
to be another RB program. As long as the data format in the
send/receive buffer is in accordance with a format that both
sockets are in agreement upon, the other socket can be:
- ➤ The same PC running another RB program.
- ➤ The same PC running a non-RB program.
- ➤ Another PC with another RB program.
- ➤ Another computer (not just a PC) running a non-RB
  program.
- ➤ A device (e.g. robot or microcontroller) that can use
  the UDP.

Sending Data Using the UDP

To send data using a UDP socket use the command
`UDP_Send(se_Name,se_Data,se_TargetIP,ne_TargetPort)`.
Notice that you must specify the destination IP and Port
number. The port number is a numeric and the IP is a string
but must be of a legal format (`"n1.n2.n3.n4"`).

Notice also that you must specify the name of the socket
you want to use to send the data. If you have started
multiple sockets in a program you will need to specify
which one to use to send the data. It does not matter which
socket you use to send data. What matters is that you
specify the IP and Port of the target socket. Nevertheless,
you need to specify the name of the socket to be used in
order to tell RB which one to use, even if you only have
one.

The parameter `se_Data` is the data buffer to be sent. See
Appendix D for how to create and manipulate the buffer.
You must specify the IP and Port number of the destination
socket every time you send data through a UDP socket.
This is because the socket does not actually establish a
connection with the destination. It does not even guarantee
that the data has arrived. The only error reporting is if the
destination IP does not exist. If the destination IP does exist
the data is sent even if the destination socket does not and
there is no way of knowing this. It is as if you are sending
mail to an apartment in a building that is not occupied. The
mail will be put through the door but there is no one there
to read it. All that the mailman cares about is that there is a
building.

UDP is a connectionless protocol in that there is no
server-client or even peer-to-peer connection. A UDP
socket can be used to send to *any* other UDP socket and it
can receive from any other UDP socket. So you can use the
one socket to send to multiple sockets.

⚠The send buffer must not exceed 2048 bytes since you cannot send more than 2048 bytes of data at a time using a UDP socket. If you need to send more than this amount do multiple sends where you divide the total bytes to be sent over multiple packets (chunks) of no more that 2048 bytes each. There is no limit on the receive buffer of a socket so it can receive more than 2048 bytes.

Receiving Data Using the UDP

When a UDP socket receives data it will be appended to the end of a receive buffer. This is performed every time a data packet is received by the socket. The buffer will continue to grow until you read it (the only limit to the size of the buffer is the memory). Once you read the buffer, the bytes in it will be returned and at the same time the buffer will be cleared.

To determine the number of bytes currently in the buffer use `UDP_BuffCount(se_Name)`. This function returns the number of bytes currently in the buffer. Use this number to determine if there are bytes in the buffer and when to read the buffer depending on the required byte count.

To obtain a socket's receive buffer use the function `UDP_Read(se_Name)`. The function will return the bytes already in the buffer and then will clear the buffer. The returned value is a string (buffer). See Appendix D for how to extract individual data fields from the string (buffer).

Checking the Socket's Status

The function `UDP_Status(se_Name)` returns a string that has information about the status of the UDP socket. The status string indicates if the socket has sent the data after a send command. It also indicates if data has been received after the socket has accepted received data. The receive status can also indicate the IP address of the originating socket. Other activities also cause the status string to change. You should check this string after utilizing the socket to ensure that the socket is actually carrying out the requested operation.

One situation where you should check the status is after a send operation. The status string will indicate if the send is successful and thus help in avoiding problems. Also the `UDP_Send()` function returns the number of bytes actually sent, so this can be another way of ascertaining any problems.

UDP does not indicate if the sent buffer has actually been received by the destination socket. If you want to ensure this you must establish some kind of *Acknowledgement Protocol* between the sockets. This means that the receiver socket must send some form of *ACK* code back to the sender for the sender to be able to determine that the data has arrived.

UDP does not have an extensive error correction protocol (unlike TCP). This means that a received buffer may have some errors that are not corrected by the rudimentary error correction mechanism that the UDP utilizes (CRC). Another shortcoming with the UDP is sequencing. If you send two packets (buffers) consecutively it is possible for the latter to arrive at the destination socket before the former. This becomes more likely if the destination is not on the same LAN where one packet takes longer to arrive due to being routed over a longer link route than the other.

Therefore you need to establish a form of Error correction and Sequence verification protocols if you wish to ensure that the data has arrived uncorrupted and is reassembled in the correct sequence. One way to minimize problems is to send short packets (fewer bytes) and have a good *ACK* handshaking protocol.

Nonetheless, over a LAN the UDP can be quite fast and error free and it is not often necessary to establish an error protocol.

Receiver-Side Automatic Header Appending

The data received by a UDP socket will be appended to its receive buffer regardless of which socket has sent the data. This means that if the socket receives data from multiple sockets there will be no way of distinguishing which data came from which socket. You can change this by using the `UDP_Header(se_Name{,true|false})` function to turn on header appending to the received data.

When header appending is activated, every packet received will have a header appended to it that specifies the number of bytes and the originator IP address. Accordingly you can parse the received data and separate the received bytes into separate buffers for each sender. This can help in separating data received from multiple senders into separate buffers for actions that require different procedures according to sender. This also can help in rejecting data from invalid senders.

Sender-Side Manual Header Appending

The receiver-side header appending can be quite powerful and useful, especially if you are communicating with systems that do not have the ability to append their own sender-side headers. Then again, a better method, which also provides for more versatility, is to have the sender append headers to the data packets. This header should have useful and application specific information. For example if multiple sockets on the same IP send data to a central socket that collects data from all the sockets, the automatic receiver-side header appending won't be able to distinguish between the sockets since they are all on the same IP and the header only holds the IP of the sender not its Port. This can be solved if the sender sockets appended a header to their data with their IP and Port and any other necessary extra bytes like the length of the packet. The length of the packet is a good thing to have especially if the packets can be of variable lengths.

Developing a UDP Program

Sending and receiving data using UDP sockets in RobotBASIC is extremely easy. To illustrate this we shall develop 4 programs. You can also see a slightly more sophisticated demo program for UDP communications in the RobotBASIC Help file in the UDP Socket Functions section.

A Simple UDP Program

The program shown in Figure C-2 is extremely simplistic as far as the user I/O is concerned and leaves much to be desired in functionality. Nevertheless, it serves to illustrate the actions of sending and receiving data through a UDP socket at its simplest without getting entangled with a GUI design or extraneous details.

The program gets a key stroke from the user then sends it through a UDP socket to another. In the program listing below the remote socket is specified to be the very same socket.

Type the program and run it. If you wish you can start two instances of RobotBASIC on the same computer and type the program in both instances but make sure that the variable **lclPort** is different in both instances. Also, make sure that the variable **rmtPort** on each instance is the value of the **lclPort** of the other instance.

```
//make sure other side is a different port number
lclPort = 46000
//change rmtIP and rmtPort to send to another socket
rmtIP = TCP_LocalIP()
rmtPort = 46000
udp_start("u1",lclPort)
while true
  if udp_BuffCount("u1") then print udp_Read("u1"),
  GetKey K
  if K
     udp_send("u1",char(K),rmtIP,rmtPort)
     waitnokey 150
  endif
wend
```

Figure C-2: This UDP program sends data to itself.

If you run the two instances on different PCs then make sure that the **rmtIP** variable in each instance reflects the IP of the other PC. To find out the IP of a machine run this one line program on it:

```
Print TCP_LocalIP()
```

> ⚠️ If you run two instances on the same PC then make sure both instances are up and running before you start typing to enter data.

That is all you need to achieve data sending and receiving simply between UDP sockets.

Another UDP Program

The program in Figure C-3 is more complex than the first one. It is still a simple program but it illustrates how you can achieve a little better user I/O.

```
lclPort = 46000
//make sure other side is a different port number
//change rmtIP and rmtPort to send to another socket
rmtIP = TCP_LocalIP()
rmtPort = 46000
udp_start("u1",lclPort)
addmemo "m1",10,10,380,400
addmemo "m2",400,10,380,400
readonlymemo "m2"
s   = "Make sure the other side is running "
s += "before you start typing."
setmemotext "m1",s
SetMemoSelection "m1",1,1,length(s)
n = memochanged("m1")
focusmemo "m1"
while true
   if udp_BuffCount("u1")
     SetMemoText "m2",udp_Read("u1")
   endif
   if memochanged("m1")
     udp_send("u1",getmemoText("m1"),rmtIP,rmtPort)
   endif
wend
```

Figure C-3: This UDP program provides a simple GUI.

> ⚠️If you run two instances on the same PC then make sure both instances are up and running before you start typing.

Just like in the previous program this one is set to send to itself. However, if you run the program in two instances of RB on the same machine or on different PCs then you need to change the variables **rmtIP** and **rmtPort** in each instance to be the values for the other instance. Also make sure the local Port numbers are different if running the two instances on the same PC.

A More Practical UDP Example

As a more practical use of the preceding information, we will develop an application that uses the UDP to send and receive textual and numerical data between two RobotBASIC programs that may run on the same machine or on separate machines within the LAN. The programs will also function between machines across the Internet, however, certain actions have to be taken with the Routers on both sides to allow for this. We shall describe these actions later in this appendix.

Each program will have one UDP socket that will serve as the sender and receiver simultaneously. Remember that each socket is assigned a port number that must be unique. Consequently, if we are to be able to run the two programs on the same machine (for easier testing) then we must ensure that the two programs will not use the same port numbers for their sockets.

The programs will intentionally be kept simple so as to facilitate understanding without too much complexity. The first program will have a user interface to obtain a text, an integer, a float and a number that is not bigger than 255 (one byte). These data will be sent to the other program that will add one to the numbers and will capitalize the text and then send it back to the sender which will display the resultant data.

We shall name the program that obtains the data from the user UDP_UserIO.Bas (Figure C-4)and the program that performs the calculations UDP_Calculator.Bas (Figure C-5).

The UDP_UserIO.Bas Program

When the program starts running it will create an edit box so that the user can enter a Port number for the socket and press a push button to activate the socket. The Port number can be defined in the edit box only upon running the program. Once the socket has been started it will remain attached to the assigned port for the duration; the edit box will be disabled and the activation button will disappear. The user must note the Local IP and Port so as to enter them in the other program's Remote IP and Port fields. Other edit boxes will then be created that will allow the user to specify the remote IP and Port number of the other program (UDP_Calculator.Bas) which can be on the same machine or another machine.

Another edit box will display the status of the socket as the program performs actions. The program will then create edit boxes to obtain the data from the user [text, two integers (one will be assumed to be less than 256) and a float]. There will also be a push button to initiate the data sending.

Once the Send button is pushed, the program will read the data from the edit boxes in order to create a buffer with the data in it that will be sent to the remote socket defined by the remote IP and Port address. Afterwards, the program will wait for data to be received. Once the data is received it will be read into a buffer from which the data will be extracted [byte, integer, float and text] and then assigned to the appropriate edit boxes for display. The entire action is repeatable as many times as the user wishes. See Appendix

D for details on what functions to use to do the data
extraction and insertion into a buffer.

```
//----------UDP_UserIO.Bas-----------
MainProgram:
  GoSub Initialization
  while true
    if LastButton() != "" then GoSub SendData
    SetEdit "Status",UDP_Status("U1")
  wend
End

Initialization:
  xyText 0,10,Center(FileNAme(ProgName()),\
                " ",45),,20,fs_Bold
  line 0,43,800,43,3
  xyText 10,50,"Local IP   = "+\
            TCP_LocalIP(),,14,fs_Bold
  xyText 10,75,"Local Port = ",,14,fs_Bold
  AddEdit "Local Port",145,75,50,0,47000
  IntegerEdit "Local Port"
  AddButton "Start Socket",230,75
  AddEdit "Status",10,110,250
  repeat //wait until user pushes the button
  until LastButton() != ""
  UDP_Start("U1",ToNumber(GetEdit("Local Port")))
  RemoveButton "Start Socket"
  EnableEdit "Local Port",false
  xyText 350,50,"Remote IP  =",,14,fs_Bold
  xyText 350,75,"Remote Port=",,14,fs_Bold
  AddEdit "Remote IP",500,50,100,0,TCP_LocalIP()
  AddEdit "Remote Port",500,75,50,0,45000
  IntegerEdit "Remote Port"
  xytext 10,200,"   Byte:        +1 =",\
              ,14,fs_Bold
  xytext 10,230,"Integer:        +1 =",\
              ,14,fs_Bold
  xytext 10,260,"  Float:        +1 =",\
              ,14,fs_Bold
  xytext 10,290,"   Text:",,14,fs_Bold
  xytext 500,260,"   Result",,14,fs_Bold
  AddEdit "Byte",100,200 \ IntegerEdit "Byte"
  AddEdit "ByteRes",270,200 \ ReadOnlyEdit "ByteRes"
  AddEdit "Integer",100,230 \ IntegerEdit "Integer"
  AddEdit "IntegerRes",270,230
  ReadOnlyEdit "IntegerRes"
  AddEdit "Float",100,260 \ FloatEdit "Float"
  AddEdit "FloatRes",270,260
  ReadOnlyEdit "FloatRes"
  AddEdit "Text",100,290,300
```

```
   AddEdit "TextRes",420,290,300
   ReadOnlyEdit "TextRes"
   AddButton "Send",420,220,100
Return

SendData:
   EnableButton "Send",false
   B = toByte(ToNumber(GetEdit("Byte"),0))
   SetEdit "Byte",Ascii(B)
   I = ToNumber(GetEdit("Integer"),0)
   F = ToNumber(GetEdit("Float"),0)*1.0
   // 1.0 ensures a float
   T = GetEdit("Text")
   s = "" \ BuffPrintB s,B,I,F,T //--create the buffer
   UDP_Read("U1")    //--clear buffer by reading it
   RIP = GetEdit("Remote IP")
   RP = ToNumber(GetEdit("Remote Port"),1)
   UDP_Send("U1",s,RIP,RP)  //--send the buffer
   repeat //--wait for at least 13 bytes (1+4+8)
      SetEdit "Status",UDP_Status("U1")
   until UDP_BuffCount("U1") >=13
   delay 100 //--allow any more bytes time to arrive
   SetEdit "Status",UDP_Status("U1")
   s = UDP_Read("U1") //--read them
   SetEdit "ByteRes", BuffReadB(s,0) // extract a byte
   // extract 4 byte integer
   SetEdit "IntegerRes",BuffReadI(s,1)
   // extract 8 byte float
   SetEdit "FloatRes",BuffReadF(s,5)
   // extract the text
   SetEDit "TextRes",BuffRead(s,13,-1)
   EnableButton "Send"
Return
```

Figure C-4: User interface side of the system.

Before data can be sent by pushing the *Send* button, the other program (UDP_Calculator.Bas) should be started and its socket activated. This is very important if you are running the two programs on the same PC (and under XP or 2000) due to the error situation discussed in the BOXES on pages 111 and 112. But it is also important if a proper response to the sent data is to be assured, whether the two programs are running on the same machine or separate machines.

ⓘThe function `TCP_LocalIP()` is used to find out what is the local IP address of the machine the program is running on. This function can be used regardless whether you are using UDP sockets or TCP sockets. It is a general network related function despite its prefix.

The UDP_Calculator.Bas Program

When the program starts running it will create an edit box for the user to assign a Port number for the socket and press a push button to activate the socket. The Port number can be defined in the edit box, but only upon running the program. Once the socket has been started it will remain attached to the assigned port number for the duration; the edit box will be disabled and the push button will disappear. The user must note the Local IP and Port so as to enter them in the other program's Remote IP and Port fields.

The program will create edit boxes to allow the user to specify the remote IP address and Port number where the other program (`UDP_UserIO.Bas`) is going to be run so as to be able to send the results back to it. It will also have another edit box to show the status of the socket for observing what is going on with the program.

Once the socket has been started the program will enter a loop waiting for data to arrive. Once data arrives, it will be read into a buffer from which the data will be extracted [the numbers and text]. The program will then perform the proper calculations and create a new buffer with the resultant data and then will send it to the remote socket as indicated by the remote IP and Port, and then go back to waiting for incoming data again. See Appendix D for details on what functions to use to do the data extraction and insertion into a buffer.

Suggested Improvements

The two programs are pretty self explanatory and work quite adequately. However, there is one shortcoming in the system. Notice both programs had a `delay 100` in order to wait for at least 13 bytes.

The reason for the delay is that we do not know the length of a packet. We know that there are at least 13 bytes (1 + 4 + 8) which are the three numbers. But the text can be of any length. So we force a wait until at least 13 bytes come in and then delay 100 ms to ensure that the bytes for the text would also have enough time to come in before we read the buffer to get the data and then act upon it.

This delay is a guesstimate. If it is too long the program is not optimal but this is not a problem. What would happen if it is too short? The buffer would be read before all the text data has had time to arrive and we would be getting the incomplete data.

There are many ways we can get around this problem. The best solution is through the use of a header. The data packet should always have an initial field that indicates the length of the data in it. We can then wait for 4 bytes. Read the buffer (it may by then have more than 4 bytes). We extract the number of bytes to be expected (say X) from the header. We then wait for (X minus the length of buffer already read) more bytes to arrive. Once they arrive we append them to the previously read data and then start manipulating the data.

```
//----------UDP_Calculator.Bas-----------
MainProgram:
  GoSub Initialization
  while true
    repeat //--wait for at least 13 bytes 1+4+8
      SetEdit "Status",UDP_Status("U1")
    until UDP_BuffCount("U1") >= 13
    delay 100 //--allow time for rest of bytes
    s = UDP_REad("U1")
    B = BuffReadB(s,0) //--extract a byte
    I = BuffReadI(s,1) //--extract integer 4 bytes
    F = BuffReadF(s,5) //--extract float 8 bytes
    T = BuffRead(s,13,-1) //--extract the rest
    xyText 120,200,B,,14,fs_Bold
    xyText 120,230,I,,14,fs_Bold
    xyText 120,260,F,,14,fs_Bold
    xyText 120,290,T,,14,fs_Bold
    s = ""  // now create the buffer
    BuffPrintB s,toByte(B+1),I+1,F+1,Upper(T)
    RIP = GetEdit("Remote IP")
    RP = ToNumber(GetEdit("Remote Port"),1)
    UDP_Send("U1",s,RIP,RP)  //--send the buffer
  wend
End

Initialization:
  xyText 0,10,Center(FileNAme(ProgName()),\
        " ",45),,20,fs_Bold
  line 0,43,800,43,3
  xyText 10,50,"Local IP   = "+\
            TCP_LocalIP(),,14,fs_Bold
  xyText 10,75,"Local Port = ",,14,fs_Bold
  AddEdit "Local Port",145,75,50,0,45000
  IntegerEdit "Local Port"
  AddEdit "Status",10,110,250
  xyText 350,50,"Remote IP  =",,14,fs_Bold
  xyText 350,75,"Remote Port=",,14,fs_Bold
  AddEdit "Remote IP",500,50,100,0,TCP_LocalIP()
  AddEdit "Remote Port",500,75,50,0,47000
  IntegerEdit "Remote Port"
  AddButton "Start Socket",230,75
  repeat //wait until user pushes the button
  until LastButton() != ""
  // now start the socket
  UDP_Start("U1",ToNumber(GetEdit("Local Port")))
  RemoveButton "Start Socket"
  enableEdit "Local Port",false
  xytext 120,170,"Received Data",,14,fs_Bold
  xytext 10,200,"   Byte:",,14,fs_Bold
  xytext 10,230,"Integer:",,14,fs_Bold
  xytext 10,260,"  Float:",,14,fs_Bold
  xytext 10,290,"   Text:",,14,fs_Bold
Return
```

Figure C-5: The calculator side of the system.

The above assures optimal waiting. There are many other ways to achieve a similar action. For example the sender can send the number of bytes to be expected and then wait for acknowledgement before it starts sending the actual data. The point is that it is up to you how to implement the **Hand-Shaking protocol** to achieve synchronized orderly and error free communications. RobotBASIC provides the link for the data, while the data content and synchronization are up to you.

Both programs use a waiting loop that waits for data to arrive at the receive buffer. This is not much of a problem in this application since the programs do not need to do much else. But, if the programs had to do other tasks while waiting for the data to arrive, then it becomes necessary to use *EVENT* handling.

RobotBASIC can be instructed to *interrupt* what it is currently doing and jump to a special routine whenever any bytes arrive at the receive buffer of *any* active UDP socket. In the *event handler* routine you would determine which active sockets have data in them and read these data and transfer them to a separate accumulator buffer for each. The handler also can initiate actions depending on which socket has received the required amount of data and so forth. Once the handler routine finishes, the program will return to doing what it was doing before the interruption.

This interrupt mechanism allows the program to do actions other than just sit in a loop waiting for data to arrive. The mechanism is activated with the statement OnUDP; read about it and a demo program in the RobotBASIC Help file.

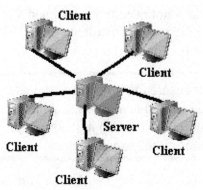

Figure C-6: Client Server Conceptual Layout

Utilizing the TCP System

A RobotBASIC program has access to *one* TCP server socket and *one* TCP client socket.

The TCP is a connection oriented Client-Server protocol. A server socket can accept multiple clients while a client socket can only be connected to one server at a time. You can disconnect a client from a server and then connect to another server; however, it can only be connected to *one server at a time*.

The server socket can accept multiple clients and will be able to receive data from all the clients and can send data to all the clients.

Once a client has connected to a server it becomes as if there is a direct two-way com link between them. Figure C-6 shows the logical conceptual layout once connections between the server and various clients have been established.

ⁱ̷The server and any of the clients can be on different LANs or the same LAN.

A server socket must be assigned a unique and not currently in use Port number. When created, the socket will *automatically* use the IP address of the machine it is created on, but, you must assign it a particular port number. The client socket does not need to be assigned a Port number. It will automatically use any available port and will use the IP of the machine it is running on.

> ⚠️You must ensure that the port assigned to a TCP Server socket is unique and is not being used by another active socket (UDP or TCP) in the same program or other programs on the same machine.

Since there is only one client socket and one server socket there is no need to name them when created or when using functions related to them

To create and activate a TCP Server socket use **TCPS_Serve({ne_Port})**. Notice that the port number is optional. If you do not specify a port number it will be assumed to be 50000. A server socket can be closed any time and reactivated with the same or even a different Port number. To deactivate a server socket use **TCPS_Close()**. No parameter is required. It is important that you monitor the status string of the server socket to know if it has been successfully activated.

Use **TCPC_Connect(se_ServerIPaddress{, ne_ServerPort})** to create and activate a TCP Client socket and at the same time connect to a server socket. Notice that the server port number is optional. If you do not specify a port number it will be assumed to be 50000, just like the server-side. This function will try to connect to the server socket as specified by the IP and Port number combination. If the server is available and running the connection will be established. A client socket can be closed any time and disconnected from the server and then reactivated to connect to a different server. To deactivate a client socket use **TCPC_Close()**. No parameter is required. It is important that you monitor the status string of the client socket to know if it has been successfully activated (see page 131).

ⓘThe client socket itself is not assigned an IP or Port number. As mentioned earlier the client socket will automatically use any available free port and the IP of the machine it is running on.

The functions discussed in the following sections enable the writing of a program to send and/or receive data packets over the TCP. A server/client socket in a RobotBASIC program can communicate over the LAN or Internet with a client/server socket that does not have to be another RB program. So long as the data format in the send/receive buffer is in accordance with a format that both sockets are in agreement upon, the other socket can be:

➤ The same PC running another RB program.
➤ The same PC running a non-RB program.
➤ Another PC with another RB program.
➤ Another computer (not just a PC) running a non-RB program.
➤ A device (e.g. robot or microcontroller) that can use the TCP.

Sending Data Using the TCP

Once a client has established a connection with a server it can send packets to it and so can the server send to the client. Use **TCPC_Send(se_Data)** to send data packets from the client to the server. Notice that you do not need to specify an IP or port the only parameter is the data buffer (string). This is of course because there is already an established link and the client can only send and receive data over this link. A client cannot send to any other server socket without disconnecting from the currently connected server and then connecting to the other server socket. A server can send data to *all* the clients that are connected to it. The server socket in RobotBASIC cannot send to a particular client only. It can send (broadcast) to all the clients connected to it. If you wish to make the data significant to only one client you need to establish a header mechanism as discussed on page 132. Use the command **TCPS_Send(se_Data)** to send data from the server socket to *all* the client sockets currently connected to the server. Notice that there is no IP or port parameter. This is because the connection link is already established and these values are already known from the link status.

In both the above functions the parameter **se_Data** is the data buffer to be sent. See Appendix D for how to create and manipulate the buffer.

⚠The send buffer in both the server and client sockets is limitless. However, from personal experience an optimal size exists. It is best if you limit the send packet to 2^{18} (262144=0x040000) bytes. This size seems to be able to reach its destination faster than smaller or bigger buffer sizes.

Receiving Data Using the TCP

When either the TCP Server socket or the TCP Client socket receives data it will be appended to the end of a receive buffer. This is performed every time a data packet is received by the socket. The buffer will continue to grow until you read it (the only limit to the size of the buffer is the memory). Once you read the buffer, the data in it will be returned and at the same time the buffer will be cleared. To determine the number of bytes currently in the buffer use **TCPC_BuffCount()** for the client socket and **TCPS_BuffCount()** for the server socket. The function returns the number of bytes currently in the buffer. Use this number to determine if there is data in the buffer and when to read the buffer depending on the required byte count. The client socket can only receive packets from the server it is connected to and thus there is no problem with knowing where the data came from. But, the server socket can receive data at any time from any of the clients it is connected to. If multiple clients exist, you must devise a way of knowing which data came from which client. Of course if you only have one client then there is no problem.

To obtain data from the sockets' receive buffers use **TCPC_Read()** for the client socket and **TCPS_Read()** for the server socket. These functions will return the bytes already in the buffer and then will clear the buffer. The returned value is a string (buffer). See Appendix D for examples of how to extract individual data fields from the string (buffer).

Checking the Socket's Status

The function **TCPC_Status()** for the client and **TCPS_Status()** for the server socket returns a string that has information about the status of the socket. The status string indicates if the socket has sent the data after a send command. It also indicates if data has been received after the socket has accepted received data.

For the client socket the status string can also indicate if the client has connected to a server and which one. It is also possible to read the status to find when and if the server has disconnected from the client. Likewise, for the server socket there are similar status strings. Read the RobotBASIC Help file for a list of these status strings and their significance.

The receive status on the server has information about the IP and Port number of the client socket that sent the data. Other activities also cause the status string to change. You should check this string after utilizing the socket to ensure that the socket is actually carrying out the requested operation.

One situation where you should check the status is after a send operation. The status string will indicate if the send is successful and thus help in avoiding problems. Another situation where you may want to monitor the status string closely is while connecting a client to a server. The status string will indicate the progress of the operation and if successful or not.

The TCP implements a very rigorous error correction and sequence assurance mechanisms. Once a link between the client and server is established you can almost always be assured that a sequence of data packets sent in succession will arrive to the other side in the correct order and will be error free and will definitely arrive. If there is any problem the status string will indicate so.

However despite all this assurance you may want to establish some kind of *Acknowledgement Protocol* between the sockets. This means that the receiver socket must send some form of *ACK* code back to the sender for the sender to be able to determine that the data has arrived. This is not strictly necessary but it may be useful for synchronization purposes in certain situations.

Server-Side Automatic Header Appending

The data received by a server socket will be appended to its receive buffer regardless of which client socket has sent the data. This means that if the socket receives data from multiple clients there will be no way of distinguishing which data came from which client. You can change this by using the **TCPS_Header({true|false})** function to turn on header appending to the received data.

When header appending is activated, every packet received will have a header appended to it that specifies the number of bytes and the originator IP address and Port number of the sending client. This way you can parse the received data and separate the received bytes into separate buffers for each client. This can help in being able to receive data from multiple clients and to put the data from each in separate storage areas for actions that require different procedures according to the client. This also can help in verifying if a client is a valid one. There is no such mechanism for the client side as there is no need for it.

Sender-Side Manual Header Appending

The server-side header appending can be quite powerful and useful, especially if you are communicating with systems that do not have the ability to append their own client-side headers. However, a better method which also provides for more versatility is to have the sender (server or client) append headers to the data packets it sends. This header should have useful and application specific information. For example a server acting as a central controller for multiple robots connected to it as clients (e.g. soccer team) can tell a particular client to do an action by appending a header to each message it sends that specifies the target client. So, despite all the clients receiving the same message, only one acts upon it, since the header indicates which client should act.

It is important to realize that the TCP functionality in RobotBASIC provides a bidirectional data link between the client and the server. It is up to you what the data content is and what methods you use for ensuring synchronization and data integrity as well as what the data content signifies.

Developing a TCP Program

Sending and receiving data using the TCP Server and TCP Client sockets in RobotBASIC is relatively easy. To illustrate this we shall develop 4 simple programs. You can also see a slightly more sophisticated demo program for TCP communications in the RobotBASIC Help file in the TCP Sockets Functions section

A Simple TCP Program

The first program is shown in Figure C-7. It is extremely simplistic as far as the user I/O and error situations detection is concerned and leaves much to be desired in functionality. Nevertheless, it serves to illustrate the actions of sending and receiving data through the TCP server and client sockets at its simplest without getting entangled with a GUI design or extraneous details.

The program gets a key stroke from the user and then sends it through the TCP Server (Client) socket to the TCP Client (Server) socket.

The program acts, both as the server and client. However, you can run two instances of the program either on the same PC or on different PCs. You must decide which instance is going to act as a Server and which will act as a Client. In the Client instance change the variables **rmtIP** and **rmtPort** to be the values of the Server's instance, and make sure the variable **IsServer** is set to false. Additionally, make sure that in the server instance the variable **IsClient** is set to false. Remember, now you will only see text on the server side when you type on the client side and vice versa. Also make sure you run the server instance BEFORE the client instance.

This program demonstrates the basic principles needed for sending and receiving data between TCP client and server sockets.

An Improved TCP Program

The second program (Figure C-8) is a little better at user I/O than the first one. As in the previous example, the program acts both as the server and client. However, you can run two instances of the program either on the same PC or on different PCs. You must decide which instance is going to act as a Server and which will act as a Client. In the Client instance change the variables **rmtIP** and **rmtPort** to be the values of the Server's instance, and make sure the variable **IsServer** is set to false. Additionally, make sure that in the server instance the variable **IsClient** is set to false. Remember, now you will only see text on the server side when you type on the client side and vice versa. Also make sure you run the server instance BEFORE the client instance.

```
//----------A Simple TCP Program----------
lclPort = 50000 //only important for the server side
//change rmtIP and rmtPort to to server's IP and Port
rmtIP = TCP_LocalIP()
rmtPort = 50000
IsServer = true
IsClient = true
if IsServer then TCPS_Serve(lclPort)
if IsClient then TCPC_Connect(rmtIP,rmtPort)
while true
  if IsServer
    if TCPS_BuffCount() then print TCPS_Read(),
    GetKey K
    if K
      TCPS_send(char(K))
      waitnokey
    endif
  endif
  if IsClient
    if TCPC_BuffCount() then print TCPC_Read(),
    GetKey K
    if K
      TCPC_send(char(K))
      waitnokey
    endif
  endif
wend
```

Figure C-7: This program illustrates the basic principles of TCP communications.

```
//-------An Improved TCP Program----------
lclPort = 50000  //important for the Server side
//--change rmtIP and rmtPort to send to another
socket
rmtIP = TCP_LocalIP()
rmtPort = 50000
IsServer = true
IsClient = true
addmemo "m1",10,10,380,400
addmemo "m2",400,10,380,400 \ readonlymemo "m2"
s   = "Start typing"
ts  = "Make sure the "
ts2 = " is running before you start typing."
if IsServer & !IsClient then s = ts+"Client"+ts2
if !IsServer & IsClient then s = ts+"Server"+ts2
setmemotext "m1",s
SetMemoSelection "m1",1,1,length(s)
```

```
n = memochanged("m1")
focusmemo "m1"
if IsServer then TCPS_Serve(lclPort)
if IsClient then TCPC_Connect(rmtIP,rmtPort)
while true
  if IsServer
    if TCPS_BuffCount()
      SetMemoText "m2",TCPS_Read()
    endif
    if memochanged("m1")
      TCPS_send(getmemoText("m1"))
    endif
  endif
  if IsClient
    if TCPC_BuffCount()
      SetMemoText "m2",TCPC_Read()
    endif
    if memochanged("m1")
      TCPC_send(getmemoText("m1"))
    endif
  endif
wend
```

Figure C-8: This TCP program offers an improved user interface.

A More Practical TCP Program

To demonstrate a more practical use of the preceding information, we will develop an application that uses the TCP to send and receive textual and numerical data between two RobotBASIC programs that may run on the same machine or on separate machines within the LAN. The programs will also function between machines across the Internet, however, certain actions have to be taken with the Routers on the server side to allow for this. We shall describe these actions later.

One program will be the server and the other will be the client. The programs will intentionally be kept simple so as to facilitate understanding without too much complexity. The client program will have a user interface to obtain a text, an integer, a float and number that is not bigger than 255 (one byte). These data will be sent to the server program that will add one to the numbers and will capitalize the text and then send it back to the client which will display the resultant data.

We shall name the client program which also obtains the data from the user `TCPC_UserIO.Bas` (Figure C-9) and the server program that also performs the calculations `TCPS_Calculator.Bas` (Figure C-10).

The TCPC_UserIO.Bas Program:

When the program in Figure C-9 starts running it will create edit boxes so that the user can enter the server's IP address and Port number (these should be obtained from the server's screen). A button will be shown that allows the user to activate the socket and connect to the server indicated by the Remote IP and Remote Port numbers. Before connecting to the server by pushing the Connect button, the server program (`TCPS_Calculator.Bas`) should be started and its socket activated. This is important if a connection is to be possible.

Another edit box will display the status of the socket as the program performs actions. The program will then create edit boxes to obtain the data from the user [text, two integers (one will be assumed to be less than 256) and a float]. There will also be a push button to initiate the data sending.

Once the Send button is pushed, the program will read the data from the edit boxes in order to create a buffer with the data in it which will then be sent to the server. The program will then wait for data to be received. Once the data is received it will be read into a buffer from which the data will be extracted [byte, integer, float and text] and then assigned to the appropriate edit boxes for display. The entire action is repeatable as many times as the user wishes. See Appendix D for details on what functions to use to do the data extraction and insertion into a buffer.

ⒾThe function `TCP_LocalIP()` is used to find out
what is the local IP address of the machine the program
is running on. This function can be used regardless
whether you are using UDP sockets or TCP sockets. It
is a general network related function despite its prefix.

```
//----------TCPC_UserIO.Bas-----------
MainProgram:
  GoSub Initialization
  while true
    if LastButton() != "" then GoSub SendData
    SetEdit "Status",TCPC_Status()
  wend
End

Initialization:
  xyText 0,10,Center(FileNAme(ProgName()),\
         " ",45),,20,fs_Bold
  line 0,43,800,43,3
  xyText 10,50,"Local IP   = "+\
         TCP_LocalIP(),"",14,fs_Bold
  AddEdit "Status",10,110,250
  xyText 350,50,"Remote IP =",,14,fs_Bold
  xyText 350,75,"Remote Port=",,14,fs_Bold
  AddEdit "Remote IP",500,50,100,0,TCP_LocalIP()
  AddEdit "Remote Port",500,75,50,0,47000
  IntegerEdit "Remote Port"
  AddButton "Connect",640,55,100
  repeat //wait until user pushes the button
  until LastButton() != ""
  RIP = GetEdit("Remote IP")
  RP = ToNumber(GetEdit("Remote Port"))
  TCPC_Connect(RIP,RP)  //--connect to server
  RemoveButton "Connect"
  xytext 10,200,"   Byte:        +1 =",\
             ,14,fs_Bold
  xytext 10,230,"Integer:        +1 =",\
             ,14,fs_Bold
  xytext 10,260,"  Float:        +1 =",\
             ,14,fs_Bold
  xytext 10,290,"   Text:",,14,fs_Bold
  xytext 500,260,"   Result",,14,fs_Bold
  AddEdit "Byte",100,200 \ IntegerEdit "Byte"
  AddEdit "ByteRes",270,200 \ ReadOnlyEdit "ByteRes"
  AddEdit "Integer",100,230 \ IntegerEdit "Integer"
  AddEdit "IntegerRes",270,230
```

```
      ReadOnlyEdit "IntegerRes"
      AddEdit "Float",100,260 \ FloatEdit "Float"
      AddEdit "FloatRes",270,260
      ReadOnlyEdit "FloatRes"
      AddEdit "Text",100,290,300
      AddEdit "TextRes",420,290,300
      ReadOnlyEdit "TextRes"
      AddButton "Send",420,220,100
Return

SendData:
      EnableButton "Send",false
      B = toByte(ToNumber(GetEdit("Byte"),0))
      SetEdit "Byte",Ascii(B)
      I = ToNumber(GetEdit("Integer"),0)
      F = ToNumber(GetEdit("Float"),0)*1.0
                        // 1.0 to ensure float
      T = GetEdit("Text")
      s = "" \ BuffPrintB s,B,I,F,T //--create the buffer
      TCPC_Read()  //--clear buffer by reading it
      TCPC_Send(s)  //--send the buffer
      repeat
          //--wait for at least 13 bytes (1+4+8)
          SetEdit "Status",UDP_Status("U1")
      until TCPC_BuffCount() >=13
      delay 100 //--allow any more bytes time to arrive
      SetEdit "Status",TCPC_Status()
      s = TCPC_Read() //--read them
      SetEdit "ByteRes", BuffReadB(s,0) //extract a byte
      //extract 4 byte integer
      SetEdit "IntegerRes",BuffReadI(s,1)
      //extract 8 byte float
      SetEdit "FloatRes",BuffReadF(s,5)
      //extract the text
      SetEDit "TextRes",BuffRead(s,13,-1)
      EnableButton "Send"
Return
```

Figure C-9: User interface side of the system.

The TCPS_Calculator.Bas Program:

When the program in Figure C-10 starts running it will create an edit box for the user to assign a Port number for the server socket and press a push button to activate the socket. The Port number of the socket can be defined in the edit box, but only upon running the program. Once the socket has been started it will remain attached to the assigned port number as long as the program is running; the edit box will be disabled and the push button will disappear. The user must note the Local IP and Port so as to enter them in the client program's Remote IP and Port fields.

```
//----------TCPS_Calculator.Bas-----------
MainProgram:
  GoSub Initialization
  while true
    repeat //--wait for at least 13 bytes 1+4+8
      SetEdit "Status",TCPS_Status()
    until TCPS_BuffCount() >=13
    delay 100  //--allow time for rest
               //of bytes to arrive
    s = TCPS_Read()
    B = BuffReadB(s,0)     //--extract a byte
    I = BuffReadI(s,1)     //--extract integer 4 bytes
    F = BuffReadF(s,5)     //--extract float 8 bytes
    T = BuffRead(s,13,-1) //--extract the text
                          //the rest of it
    xyText 120,200,B,,14,fs_Bold
    xyText 120,230,I,,14,fs_Bold
    xyText 120,260,F,,14,fs_Bold
    xyText 120,290,T,,14,fs_Bold
    s = ""
    BuffPrintB s,toByte(B+1),I+1,F+1,Upper(T)
    TCPS_Send(s)  //--send the buffer (s)
  wend
End

Initialization:
  xyText 0,10,Center(FileNAme(ProgName()),\
            " ",45),,20,fs_Bold
  line 0,43,800,43,3
  xyText 10,50,"Local IP   = "+\
               TCP_LocalIP(),,14,fs_Bold
  xyText 10,75,"Local Port = ",,14,fs_Bold
  AddEdit "Local Port",145,75,50,0,47000
  IntegerEdit "Local Port"
  AddEdit "Status",10,110,250
  AddButton "Serve",230,75
  repeat //wait until user pushes the button
  until LastButton() != ""
  //start server
  TCPS_Serve(ToNumber(GetEdit("Local Port")))
  RemoveButton "Serve"
```

```
enableEdit "Local Port",false
xytext 120,170,"Received Data","",14,fs_Bold
xytext 10,200,"   Byte:",,14,fs_Bold
xytext 10,230,"Integer:",,14,fs_Bold
xytext 10,260,"  Float:",,14,fs_Bold
xytext 10,290,"   Text:",,14,fs_Bold
Return
```

Figure C-10: The calculator side of the system.

The program will also create an edit box to show the status of the socket for observing what is going on with the program. Also any received data will be displayed on the screen.

Once the socket has been started the program will enter a loop waiting for data to arrive. A client can connect to the server then start sending data to it. Once data arrives, it will be read into a buffer from which the data will be extracted [the numbers and text]. The program will then perform the proper calculations and create a new buffer with the resultant data and then will send it to the clients connected to it, and then go back to waiting for incoming data again. See Appendix D for details on what functions to use to do the data extraction and insertion into a buffer.

Suggested Improvements

The two TCP programs just discussed are pretty self explanatory and work quite adequately. However, there is one shortcoming in the system. Notice that each program used delays of 100 ms to wait for at least 13 bytes.

The reason for the delay is that we do not know the length of a packet. We know that here are at least 13 bytes $(1 + 4 + 8)$ which are the three numbers. But the text can be of any length. So we force a wait until at least 13 bytes come in and then delay 100 ms to ensure that the bytes for the text would also have enough time to come in before we read the buffer to get the data and then act upon it. This delay is a guesstimate. If it is too long the program is not optimal but this is not a problem. What would happen if it is too short? The buffer would be read before all the text data has had time to arrive and we would be getting the incomplete data.

There are many ways we can get around this problem. The best solution is through the use of a header. The data packet should always have an initial field that indicates the length of the data in it. We can then wait for 4 bytes. Read the buffer (it may by then have more than 4 bytes). We extract the number of bytes to be expected (say X) from the header. We then wait for (X minus the length of buffer already read) more bytes to arrive. Once they arrive we append them to the previously read data and then start manipulating the data.

The above assures optimal waiting. There are many other ways to achieve a similar action. For example the sender can send the number of bytes to be expected and then wait for acknowledgement before it starts sending the actual data. The point is that it is up to you how to implement the *Hand-Shaking protocol* to achieve synchronized orderly and error free communications. RobotBASIC provides the link for the data, while the data content and synchronization are up to you.

RobotBASIC provides the link for the data, but the data content and the data sequencing is all up to you.

Both programs use a waiting loop that waits for data to arrive at the receive buffer. This is not much of a problem in this application since the programs do not need to do much else. But, if the programs had to do other tasks while waiting for the data to arrive, then it becomes necessary to use *EVENT* handling.

RobotBASIC can be instructed to *interrupt* what it is currently doing and jump to a special routine whenever any bytes arrive at the receive buffer of a TCP socket. In the routine (*event handler*) you would read the data and transfer it to a separate accumulator buffer. The handler also can initiate actions depending on if the socket has received the required amount of data and so forth. Once the handler routine finishes, the program will return to doing what it was doing before the interruption.

This interrupt mechanism allows the program to do actions other than just sit in a loop waiting for data to arrive. The mechanism is activated with the statements **OnTCPS** and **OnTCPC**; read about them in the RobotBASIC help file.

Selecting a Port Number

When activating a TCP Server or when creating and starting a UDP socket the IP address for the socket is determined by the machine it is running on. But the Port number has to be assigned.

Either you as the programmer can hard code the Port number in your program, or you can provide a means for the user of your program to select a port number. Regardless of which option you opt for a Port number has to be selected and assigned to the socket.

How would you select a number? Well the easy answer from experience is that it is safer mostly to choose numbers greater than 40000 but any port that your system is not currently using will suffice.

The longer answer is that there are many ports that are universally agreed upon as standard ports for usage with ubiquitous programs. For example FTP is usually assigned to port 21, SMTP to port 25, HTML to port 80 and so forth. There are ports exclusively for UDP and others for TCP. Many other applications assign port numbers for UDP usage and TCP usage.

So try to avoid these ports. To help you in deciding what ports are likely to be used on your system and therefore to avoid, see these two web pages in order of preference:

> http://www.iss.net/security_center/advice/Exploits/
> Ports/default.htm
> http://www.portforward.com/cports.htm

It is generally true that ports above 40000 are safe to use.

⚠️Do not ever use ports that are assigned for HTML, SMTP, FTP and other universal Internet services.

Allowing Internet Throughput

All the information given here is equally applicable whether you are running your application within the LAN or across the Internet. As long as you have the right IP address and the right Port number for the target UDP socket you can send data to it, and if you have the right IP address and Port number of a TCP server socket a client socket can connect to it.

With UDP sockets you need the IP and Port number for both ends to allow bidirectional communications. With TCP you need to know the IP and Port of the server, the client can just connect and the link is then automatically bidirectional without ever knowing the IP address or Port number of the client. However you may want to know the IP addresses and Port numbers of valid clients so as to be able to restrict access to just valid clients.

When it comes to communicating to a machine outside your LAN you are going to have to deal with the Fire Wall and NAT aspects of the system. An IP address within the LAN is not a valid address for usage by a machine outside the LAN. This means that even though your machines local IP address is known it cannot be used as a remote IP by another PC outside your LAN. It is a usable and legal IP address for machines within the LAN but not across the Internet.

If you look back at Figure C-1 you will notice that the routers have two distinct addresses each. The addresses that start with 192.168 are known as the inside the LAN address. This is the address you can connect to your router by a machine inside the LAN. Also notice that all the machines inside the LANs have addresses that start with

192.168 and that LAN1 machines have the same addresses as LAN2 machines. Normally it is not legal for two machines to have the same address on the Internet. Each machine that is visible to the global Internet must have a unique IP address, but since the machines inside the LAN are not directly visible to the Internet, it is possible for this address duplication to occur.

The IP standard has set aside an address space that always starts with 192.168 which is designed to be a LAN group of addresses and that the global Internet will not use or assign to actual real machines. This is what the NAT does. It translates LAN addresses so as to appear as if they are valid addresses on the Internet. How this is achieved does not concern us. But what concerns us is that we need to devise a method for a machine on LAN2 to communicate with a machine on LAN1.

As far as the Internet is concerned your entire LAN is one machine given one address that is usually assigned by the ISP. In Figure C-1 you can see that LAN1 is given the address 67.134.100.225. This is the address of your machine as far as any machine outside the internet is concerned. This IP is assigned by the ISP usually on a lease basis. That means that it is not a permanent address and if your LAN ever disconnects from the ISP and then reconnects you will end up being assigned another IP that, more often than not, is different. It is possible to have a permanent (static) address given to you but this is normally only possible for bigger companies that have big LANs. The final outcome is that your LAN as far as the Internet is concerned is one machine with one IP address. In the setup in Figure C-1 the router is what the Internet sees and nothing else. So any machine trying to log on to your system can only go through the router.

⚠️If your network setup differs from Figure C-1 you will need to consult a network administrator on how to setup your system to allow access to a machine within the LAN from a machine outside the LAN. This discussion may help you get an idea of what is required in general but will <u>not</u> be applicable due to the difference.

You can use the following web site to ascertain what your LAN's IP address is. But remember this can change if you ever disconnect your system from the ISP.

http://www.whatismyip.com/

Another way you can get your Global (External) IP is by logging on to your router, which is also an operation that will be necessary for accomplishing the task of setting up your LAN so that an outside machine can connect with a machine inside the LAN. This action will be described shortly.

Once we know what IP address to give to the external machine to use in the UDP or TCP functions that require them, we also have decided on a port to use as described on page 143 and we have all we need. A machine say in LAN2 can now communicate with a machine in LAN1 using the IP address let's say 67.134.100.225, and the chosen port, let's say 50000.

Well, not quite! There is an additional problem we need to resolve. The IP address is not in fact the address of any PC inside LAN1; it is really the address of the router. *So how do we get the data packets that will reach the router to go to the PC that in fact is running the destination socket?*

> ⓘ The answer the above question is something called
> *Port Forwarding*. Simply, port forwarding is a way to
> tell the router that whenever it sees a data packet
> coming to it on a particular port to forward that packet
> to a PC inside the LAN. This PC will see the data
> packet and it will appear to it as if it came directly from
> the machine on LAN2.

To summarize the steps so far are:

1- Decide what type of socket you are going to use, TCP
   Server or UDP (e.g. UDP)
2- Decide what Port your socket is going to be
   associated with (e,g. 45000)
3- Find out what is the internal LAN address of the
   machine hosting the socket (e.g 192.168.0.110). You
   can use the `TCP_LocalIP()` function to do this.
4- Log on to your router and set it up to do port
   forwarding of Port 45000 for UDP traffic to machine
   192.168.0.110.
5- Reboot your router. This is needed for the changes to
   take effect. It will also change you external IP address
   so there is no need to have found it before this step.
6- Find out what your External IP address is. You can do
   this by going to the web site above or by logging onto
   the router again and reading the information. Say
   87.134.100.225
7- Give this external IP address and the port number
   (45000) to the machine on LAN2 that is going to send
   data to your UDP socket.
8- Start your program on the assigned machine and wait
   for data to come in.
9- Start the machine inside LAN2 and start a UDP
   socket and tell it to send data to the remote IP
   87.134.100.225 and remote Port 45000.
10- That is all.

⚓For TCP you only need to set up port forwarding for the server side. A client, once connected to a server, will establish a bidirectional link automatically and there is no need to set up port forwarding on the client side.

⚠For UDP you need to setup port forwarding on each side that is going to receive data.

Configuring the Router for Port Forwarding

The following procedure will show how to set up port forwarding for a very specific router (the one I have). Your router will have a different layout. However, the procedure is generally similar and other than different menus or screen layouts the principles and most of the terminologies are the same on most routers. If you need help with your router ask technical support from your vendor or visit this website:

http://portforward.com/

To access your router use your Internet Browser and type in the URL space the IP address of your router. Most of the time this is 192.168.0.1 but it may be 0.0 or 1.0 or 0.2 etc. Consult the router's manual or help desk.

You will then see a logon screen on your browser, perhaps similar to the one shown in Figure C-11. You will have to login using a password. If you do not know it, ask your technical support.

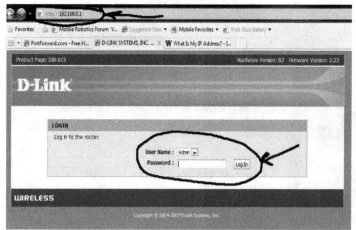

Figure C-11: Typical Router Logon Screen.

After logging in, you should see a screen like the one shown in Figure C-12. Notice the External IP address information. But this is going to change later so ignore it for now. Click on the advanced menu option.

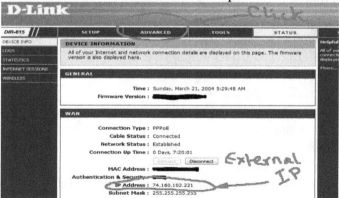

Figure C-12: A typical router screen.

When you have navigated to the screen in Figure C-13, fill in the indicated areas:

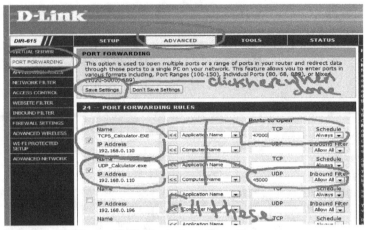

Figure C-13: Typical configuration screen.

After you finish click 'Save Settings' which will cause the router to reboot and will also cause it be assigned a new external IP which you can find out by going back the 1st screen (you have to log in again) or by going to the web site mentioned earlier.

That is all! You now have Port Forwarding. As far as the internet is concerned the machine on LAN1 with the IP address 192.168.0.110 will appear to the outside world as IP 87.134.100.225 for any TCP traffic to Port 47000 and UDP traffic to Port 45000.

In the case of the UDP example we developed earlier in this Appendix you will also have to do the same for the machine on LAN2. In the case of TCP you do not have to do port forwarding on the other machine since it will be the client and thus will have automatic two way traffic.

> ⓘThe above is a very specific example for a specific network layout and specific router. If yours is a different setup or router you will have to take the above as a general guide. While the specific actions may vary, the principles are the same. You need to establish a port forwarding system.

⚠️Port forwarding is a hole in your fire wall. While it is not much use to any hackers you may want to experiment with a machine that does not have any sensitive data.

Appendix D

Byte Buffers in RobotBASIC

This appendix covers some advanced programming methods available in RobotBASIC to do with serial and TCP/UDP I/O. Many of the operations can be handled quite effectively without using these techniques, but they can be an advantage when doing more advanced actions.

Manipulating A Byte Buffer

RobotBASIC has specific commands that can increase the effectiveness of TCP and UDP communication. The information in this section is also helpful when performing advanced Serial I/O and Low-Level File I/O as well as for utilizing the suite of USBmicro functions to control the U4x1 USB I/O devices.

In functions that send and receive data using the TCP, UDP, or Serial I/O you are in fact sending and receiving a byte buffer that contains the data to be transferred. This buffer can be manipulated in one of two ways:

1- **As a string** which can be manipulated with numerous string functions. There are functions to extract parts of the string, extract a particular character, extract a particular character as a byte, and insert characters or bytes.

> ⓘAll string functions index characters starting with one. That is the first character is index position 1.

2- **As a byte array** which can be managed with a set of specialized functions and commands to insert in it or extract from it bytes, integers, floats, or text.

> ⓘAll buffer functions index bytes starting with zero. That is the first byte is index position 0.

> ⓘWe shall refer to the buffer sometimes as a *string buffer* and at other times as a *byte buffer*. It does not matter how it is referred to, it is an *array of bytes*. The individual bytes can be ASCII characters (actual text) or they can be binary values.

> ⓘAs the programmer all you are interested in is how to extract data from the buffer when you receive it and how to insert data into the buffer in preparation for sending it. To do these operations RobotBASIC provides a suite of functions and commands for treating the buffer as an array of bytes. Additionally all the string functions in RB can be used to handle the buffer as a string.

Putting Text and Textual Numbers in a String Buffer

The buffer is in reality a string variable that can be handled in all the normal ways you treat any normal string variables. So for example if you want to have a buffer that consists of the text "Hello there" then you would do:

```
SB = "Hello there"
```

The buffer becomes the variable SB, and it now contains 11 bytes which have the values:

```
72 101 108 108 111 32 116 104 101 114 101
 H   e   l   l   o     t   h   e   r   e
```

If you later say:

```
SB = SB+" "+ (-45.7)
//conversion is performed implicitly
```

Or you can say

```
SB = SB+" "+ToString(-45.7)
//conversion is performed explicitly
```

The buffer will now be holding 17 bytes with the additional 6 bytes being:

```
32 45 52 53 46 55
    -  4  5  .  7
```

Notice how the number has been converted to its textual representation and that the characters in the string are the ASCII codes for the individual digits of the number including the minus sign and the decimal point.

You can build a string buffer this way with as many expressions as you need over as many lines of code as required. Once the variable **SB** is populated with the data it can then be used as the parameter for the **UDP_Send()**, **TCPS_Send()** or **TCPC_Send()** or **SerialOut** or **SerOut** functions which send the bytes in the string over the corresponding protocol.

A command you may find useful for populating the string (buffer) with text or textual numbers is the **BuffPrintT** command. This command is very much like the **Print** command except it will put the resulting text of the expressions in a specified string (buffer). Numbers will be put as their textual representations and you can even use semicolons and commas to control tab spacing just like you would with the **Print** command.

For example:
```
X = 3.5e200
BuffPrintT SB,X," * ",2," +5 = ",2*X+5,
BuffPrintT SB,sRepeat(" ",3);"OK",
```

Notice that with the first usage of **BuffPrintT**, **SB** does not exist and therefore will be created as an empty string. With the second usage, **SB** already exists and has data in it, so the new data will be appended to it.

> ⓘRemember this when using the command. If **SB** already exists and you do not wish to append to it you will need to say **SB = ""** to make it into a blank string before you use it with the command.

After the above sequence of code SB will contain:

51	46	53	69	50	48	48	32	42	32	50	32	43	53	32	61
3	.	5	E	2	0	0		*		2		+	5		=

32	55	69	50	48	48	32	32	32	32	32	32	32	32	32	32
	7	E	2	0	0										

79	75
O	K

Notice the comma at the end of each line in the program. This is necessary if you do not want a CR/LF [char(13)+char(10)] to be part of the buffer **SB**. The following lines:
```
X = 3.5e200
BuffPrintT SB,X," * ",2," +5 = ",2*X+5
BuffPrintT SB,sRepeat(" ",3),"OK"
```

will cause SB to have the following:

51	46	53	69	50	48	48	32	42	32	50	32	43	53	32	61
3	.	5	E	2	0	0		*		2		+	5		=

32	55	69	50	48	48	13	10	32	32	32	79	75	13	10
	7	E	2	0	0						O	K		

Notice the characters with the byte values 13 and 10 right after 200 and after the OK. These are the Carriage Return and Line Feed character pairs that normally result in a new line. Since **BuffPrintT** behaves just like a **Print** then these characters will be there if you do not end the command with a comma to stop it from inserting a CR/LF character pair.

Putting Text and Binary Numbers in a Buffer

As far as text is concerned all the details in the previous section apply. As far as binary numbers there are three situations:

1- Byte numbers, which are numbers that range from 0 to 255.
2- Integers, which are 4 bytes long (in RB) and range from **MinInteger()** to **MaxInteger()**
3- Floats are 8 bytes long (in RB) and can range from +/-**MinFloat()** to +/-**MaxFloat()**.

Bytes

RobotBASIC does not have a byte type. Nonetheless, an integer can be truncated to become a byte. There are two functions that can convert an integer to a byte value that will be usable for adding to a string (buffer). These are **Char()** and **toByte()**. They do the exact same job but are named so as to be appropriately self-documenting for whatever situation requires the function.

Both functions in reality return a buffer (string) of one byte (character). The byte will have the value of the *Least Significant Byte (LSByte)* of its integer parameter. This is the first byte from the right if you represent the number as binary (or hex).

For example if you say **SB = char(156)** or **SB = toByte(156)** then **SB** will become a one character string (one byte buffer) with the byte having the value 159. However if you say **SB = char(456)** or **SB = toByte(456)** **SB** will become one character string (one byte buffer) with the byte value being 200. Why 200? Because if you look at the hex equivalent of 456 (=0x01C8) you will see that it is 2 bytes long and the LSByte is 0xC8 (=12*16+8 = 200). For example if you want to create a buffer with the text "Hello" and the byte numbers 34 and 211, you would write:

```
SB = "Hello"+char(34)+char(211)
```

or

```
SB = "Hello"+toByte(34)+toByte(211)
```

SB will then contain the following bytes (notice the last two bytes are the numbers specified):

72	101	108	108	111	34	211
H	e	l	l	o		

Another function that can be more convenient in certain instance is **PutStrByte()**. With this function you can insert a byte value at a specific position in a string (buffer). For example:

```
SB = "Hello"
SB = PutStrByte(SB,Length(SB)+1,34)
SB = PutStrByte(SB,Length(SB)+1,211)
```

The above code will result in **SB** being exactly as listed above. This looks more complicated in this situation; the previous method is more convenient. However, if you read the details of this function in the RobotBASIC Help file, you will see that it facilitates certain actions that are required in certain situations and is a good function to remember when the need arises. Its converse **GetStrByte()** is more frequently needed.

BuffWriteB() is another function that is very similar to **PutStrByte()** but treats the buffer as a byte array and thus indexes using 0 as the first byte and so forth. The following example will result in SB being exactly as above:

```
SB = "Hello"
SB = BuffWriteB(SB,-1,34)
//-1 means the end of the existing buffer
SB = BuffWriteB(SB,-1,211)
```

Read about these two functions in the RobotBASIC Help file. They are useful and once you know how they work you will see that they are needed and necessary in certain situations that you may not appreciate from the simple example given above.

Integers and Floats

How floats and integers are represented depends on the computer you are using. RobotBASIC runs under the Windows operating system which usually runs on a PC that has an Intel processor. Integers supported by RB are 4 bytes long and are stored in what is called the *Little-Endian* format. This format stores the 4 bytes of an integer in the order from left to right with the LSByte as the first byte.

Therefore, a number like 0xA412B8D7 will be stored as D7, B8, 12, A4. Notice it is reverse to the way we normally look at binary numbers. This is just the way it is and you just have to accept it (actually it makes sense if you consider the low level data transfer mechanisms within the processor).

Some processors (e.g. 68HC11) store integers in the *Big-Endian* format (reverse) and if you are going to send numbers to devices based on these processors then it makes a difference what format is used. However, if you are going to be sending buffers between machines using the Little-Endian format you do not have to be concerned with how the numbers are stored, RB takes care of that.

With floats they are stored in a format called the IEEE 754 standard. This can get quite complex and shall not be discussed here. Just know that a float in RB is stored as 8 bytes long, what these bytes are and what values and order and so forth is immaterial. RB will take care of it.

The command **BuffPrintB** is one way to create a buffer with binary integers and floats in it. For example:

```
X = 0x00A243C1
BuffPrintB SB,X,"*",2,"+5=",2*X+5,"OK"
```

This will result in the variable SB holding the following
bytes:

193	67	162	0	42	2	0	0	0	43	53	61	135
C1	43	A2	00	*	02	00	00	00	+	5	=	87

135	68	1	79	75
87	44	01	O	K

> ⓘNotice how there is no CR/LF after OK even
> though there was no comma at the end. This is
> because **BuffPrintB** does not behave quite like the
> **Print** command. It is a binary formatter and will not
> therefore insert CR/LF as is required with text. Also
> there is no tabbing. A semicolon will have no effect.
> If you wish to insert a CR/LF use the **CrLf()**
> function.

> ⓘNotice how the numeric value 2 was inserted as an
> integer (4 bytes) even though it can fit in a byte. This
> is because the **BuffPrintB** command will not
> make assumptions. If you wish to treat an integer as a
> byte you need to convert it to one.

Notice the bold area in the following code:

```
X = 0x00A243C1
BuffPrintB SB,X,"*",toByte(2),"+5=",2*X+5,"OK"
```

In this case there will only be one byte for the number 2.

The functions **BuffWrite()** and **BuffWriteB()**
are also of use. The following example will result in SB
holding the same bytes as the example just above (also the
next section below).

```
X  = 0x00A243C1
SB = BuffWrite("",0,X)
SB = BuffWrite(SB, \
        -1,"*")+BuffWriteB("",0,2)+"+5="
SB = BuffWrite(SB,-1,2*X+5)+"OK"
```

Extracting Text and Numbers from a Buffer

When you receive a buffer with data in it you must know in what arrangement the data is organized. There has to be an agreement between the sender and receiver so that data can be inserted and extracted in the correct manner. This is especially important when there is a mixture of number types and text. This is best illustrated with a concrete example.

We will create a buffer that is a *record* of data. The record will be divided into *fields*.

Let's say you are sending a database with data about people. Each record will be transmitted as one buffer. In each of our hypothetical records there are the following fields:

Code, Name, Address, Zip_Code, Balance

In our example the Code will be considered to be a byte, the Zip_Code will be an integer and the Balance a float. We must also decide the following:

➢ Will the numbers be stored in the record as strings or as binaries?
➢ Will the text fields be of fixed lengths or variable lengths?
➢ If the text fields are to be of variable lengths how do we know where they end?

These questions have to be answered carefully in order to be able to store the data in the buffer to be transmitted, but even more crucially, so as to be able to extract the values for each field correctly.

Let's say we used the following code to create the buffer **SB** to hold the record:

```
Code    = 1
Name    = "Sam"
Address = "Here and there"
Zip_Code= 55667
Balance = 100.23
SB = ToString(Code)+Name+Address+  \
                Zip_Code+Balance
```

Will result in SB holding

```
1SamHere and there55667100.23
```

There is no problem at all in creating the buffer. The problem, though, arises when we try to extract the data from the buffer. Where does the Name field start and end? Where does the zip code field start and end? As you can see we have not created the buffer in a good way. This code is a lot better:

```
BuffPrintT SB,Code,"|",Name,"|",Address,"|",\
        Zip_Code,"|",Balance,
```

Will result in **SB** holding

```
1|Sam|Here and there|55667|100.23
```

With a record like this it is very easy to extract the various fields using RB's **Extract()** function:

```
Code     = toNumber(Extract(SB,"|",1),0)
    //defaults to 0 if bad text
Name     = Extract(SB,"|",2)
Address  = Extract(SB,"|",3)
Zip_Code = ToNumber(Extract(SB,"|",4),0)
Balance  = ToNumber(Extract(SB,"|",5),0.0)
```

Notice how we had to convert the zip code and balance to a numbers from the text. Also notice that the delimiter character has to be chosen with care. The character must not be likely to occur as part of the bytes of the fields.

Another Design

With the above scheme we stored the numbers in the buffer as text. This can be wasteful. For instance, if Code is 234 it will occupy 3 bytes ('2', '3', and '4'). Conversely, if we store it as a byte value, it will only be one byte. Also notice the zip code; it is 5 bytes long, but if we store it as an integer we would save one byte. Additionally, the balance field is a float. Imagine if Sam was a rich guy and had 10,000,000.23 in his account (wishful thinking). That would require 11 bytes to store as a text (no commas). You can see that it is better to use 8 bytes to store the float as a binary rather than text.

If we store numbers as binary there will be no possible delimiter character to use to delimit where the text fields end since binary numbers could be any values and there would be no way of having a byte value that cannot occur as part of a field. The problem is not with the numbers since we know their lengths; it is the text fields that pose a problem.

A possible solution is to fix the length of the text. So we would say that Name cannot be longer than 20 characters (bytes) and if it is less than 20 it will be padded with spaces (see **JustifyL()**). However, this is wasteful and limiting. If the name is a lot shorter than 20 characters then we would be storing too many unnecessary characters. If it needs to be longer than 20 characters then we have deteriorated the flexibility of the application.

An Efficient and Flexible Design

A better solution is to store a number before each text field that specifies the length of the text to follow. If we know that the text cannot be any longer than 255 character we can store the number as a byte. If 255 is too short then we can store the number as an integer (4 bytes). Either way is a good method and will result in the most efficient usage of the buffer. These program lines create the buffer to be sent.

```
Code     = 1
Name     = "Sam"
Address = "Here"
Zip_Code= 55667
Balance = 100.23
BuffPrintB SB,toByte(Code),Length(Name),Name
BuffPrintB SB,Length(Address), \
               Address,Zip_Code,Balance
```

The obvious question is how do we extract the individual fields from the buffer? Remember, we know that just before every text field there is an integer that indicates how long the text that follows is. Here is how we can extract the fields' values.

```
X = 0
Code=BuffReadB(SB,X)        \ X= X+1
n=BuffReadI(SB,X)           \ X=X+BytesCount_I
Name=BuffRead(SB,X,n)       \ X=X+n
n=BuffReadI(SB,X)           \ X=X+BytesCount_I
Address=BuffRead(SB,X,n)    \ X=X+n
Zip_Code=BuffReadI(SB,X)    \ X=X+BytesCount_I
Balance=BuffReadF(SB,X)     \ X=X+BytesCount_F
```

We are able to specify the exact positions for reading the various fields from the array of bytes (buffer) using the appropriate **BuffRead/B/I/F()** function. Notice a postfix of I means integer, a postfix of F means float, B means byte and no postfix means text. Notice that with the text form of the function we must also specify the number of bytes to be read. This is where the previously extracted integer that specifies the length comes in use. Also notice how we keep track of the next position within the buffer to read from using the counter X. The constants **BytesCount_I** and **BytesCount_F** are defined in RobotBASIC as 4 and 8 respectively to specify the lengths of an integer and a float in bytes. So you won't have to fix these values in case there are future modifications to the internal representation of integers and floats within RB.

Index

bit, 1
bit-wise operations, Appendix A
 clearing bits, 101
 ignoring bits, 100
 inverting bits, 102
 setting bits, 101
 toggling bits, 101
BS2 (Parallax), 19-24
buffering, 32
by-reference (passing variables), 29
byte, 2
byte buffer (advanced programming methods), Appendix D
Enhancing the Pololu 3pi with RobotBASIC, 89
gear-head motor, 31-32
GUI, 35-37
H-bridge, 33
Hardware Interfacing & Control Protocol, 95
I/O Ports, 1-6
 buffering, 32
 parallel, 1, 7-18
  printer, 8-11
  USBmicro, 5, 13-18
  virtual, 5, 25
   input, 25-26
   output 26-27
 Serial, 2, 19-30
  asynchronous, 2-3
  finding port numbers, Appendix B
  I^2C, 4
  RS-232, 4
  Synchronous, 2-3
  USB, 13
  with microcontrollers, 19-24
  virtual, 4

Internet communication and control, 81-86
  controlling a simulated robot, 82-86
motor control, 31-42
  GUI, 35-37
  servomotors, 37-38
    Micro Maestro 6 Channel USB Servo Controller, 38-39
  speed, 32-33
  Robotclaw 2x5A (Lynxmotion), 33-36
new paradigm, 88-94
  internal protocol (RobotBASIC), 91-94
Pololu 3pi, 46
pull-up resistor, 11
Robot Programmer's Bonanza, 88
sensors, 43-53
  bumper, 44
  compass, 51
  IR, 44-45
  line sensors, 51
  other, 53
  ultrasonic ranging, 46-50
servomotors, 37-38
speech, 55-63
  recognition, 56-59
    an application, 59-63
  synthesis, 55-56
TCP communication, Appendix C
vision, 65-80
  RoboRealm, 65, 72-
  web cams, 65
  with RobotBASIC, 66-72
  with RoboRealm, 72-80
UDP communication, 79, Appendix C
USB2SER (Parallax), 24-25

www.ingramcontent.com/pod-product-compliance
Lightning Source LLC
Chambersburg PA
CBHW071154050326
40689CB00011B/2114